BANACH ALGEBRAS

Modern Analytic *and* Computational Methods *in* Science *and* Mathematics

A GROUP OF MONOGRAPHS AND ADVANCED TEXTBOOKS

Richard Bellman, EDITOR

Published

1. R. E. Bellman, R. E. Kalaba, and Marcia C. Prestrud, Invariant Imbedding and Radiative Transfer in Slabs of Finite Thickness, 1963
2. R. E. Bellman, Harriet H. Kagiwada, R. E. Kalaba, and Marcia C. Prestrud, Invariant Imbedding and Time-Dependent Transport Processes, 1964
3. R. E. Bellman and R. E. Kalaba, Quasilinearization and Nonlinear Boundary-Value Problems, 1965
4. R. E. Bellman, R. E. Kalaba, and Jo Ann Lockett, Numerical Inversion of the Laplace Transform: Applications to Biology, Economics, Engineering, and Physics, 1966
5. S. G. Mikhlin and K. L. Smolitskiy, Approximate Methods for Solution of Differential and Integral Equations, 1967
6. R. N. Adams and E. D. Denman, Wave Propagation and Turbulent Media, 1966
7. R. L. Stratonovich, Conditional Markov Processes and Their Application to the Theory of Optimal Control, 1968
8. A. G. Ivakhnenko and V. G. Lapa, Cybernetics and Forecasting Techniques, 1967
9. G. A. Chebotarev, Analytical and Numerical Methods of Celestial Mechanics, 1967
10. S. F. Feshchenko, N. I. Shkil', and L. D. Nikolenko, Asymptotic Methods in the Theory of Linear Differential Equations, 1967
11. A. G. Butkovskiy, Distributed Control Systems, 1969
12. R. E. Larson, State Increment Dynamic Programming, 1968
13. J. Kowalik and M. R. Osborne, Methods for Unconstrained Optimization Problems, 1968
14. S. J. Yakowitz, Mathematics of Adaptive Control Processes, 1969
15. S. K. Srinivasan, Stochastic Theory and Cascade Processes, 1969
16. D. U. von Rosenberg, Methods for the Numerical Solution of Partial Differential Equations, 1969
17. R. B. Banerji, Theory of Problems Solving: An Approach to Artificial Intelligence, 1969
18. R. Lattès and J.-L. Lions, The Method of Quasi-Reversibility: Applications to Partial Differential Equations. Translated from the French edition and edited by Richard Bellman, 1969
19. D. G. B. Edelen, Nonlocal Variations and Local Invariance of Fields, 1969
20. J. R. Radbill and G. A. McCue, Quasilinearization and Nonlinear Problems in Fluid and Orbital Mechanics, 1970
26. W. Sierpinski, 250 Problems in Elementary Number Theory, 1970

In Preparation

21. W. Squire, Integration for Engineers and Scientists
22. T. Pathasarathy and T. E. S. Raghavan, Some Topics in Two-Person Games
23. T. Hacker, Flight Stability and Control
24. D. H. Jacobson and D. Q. Mayne, Differential Dynamic Processes
25. H. Mine and S. Osaki, Markovian Decision Processes
27. E. D. Denman, Coupled Modes in Plasms, Elastic Media, and Parametric Amplifiers
28. F. A. Northover, Applied Diffraction Theory
29. G. A. Phillipson, Identification of Distributed Systems
30. D. H. Moore, Heaviside Operational Calculus: An Elementary Foundation

BANACH ALGEBRAS

by

WIESŁAW ŻELAZKO

ELSEVIER PUBLISHING COMPANY
AMSTERDAM—LONDON—NEW YORK

PWN—POLISH SCIENTIFIC PUBLISHERS
WARSZAWA

1973

Published in co-edition with
PWN—POLISH SCIENTIFIC PUBLISHERS
WARSZAWA

Distribution of this book is being handled by the following publishers

for the U.S.A. and Canada

AMERICAN ELSEVIER PUBLISHING COMPANY, INC.
52 Vanderbilt Avenue, New York, N.Y. 10017

for Albania, Bulgaria, Chinese People's Republic, Hungary, Czechoslovakia, Cuba, German Democratic Republic, Hungary, Korean People's Democratic Republic, Mongolia, Poland, Rumania, Democratic Republic of Viet Nam, U.S.S.R. and Yugoslavia

PWN—POLISH SCIENTIFIC PUBLISHERS
Warszawa (Poland), Miodowa 10

for all remaining areas

ELSEVIER PUBLISHING COMPANY
335 Jan Van Galenstraat, P.O. Box 211
Amsterdam, The Netherlands

ISBN 0-444-40991-2

Library of Congress Catalog Card Number 72-87969

PRINTED IN POLAND

Contents

ERRATA

Page, line	For	Read
IX^{14}	W^{x} algebras	W^{*} algebras
34^{8}	Żelazko [7]	Żelazko [8]
$68_{15,\,25}$	$x \in \operatorname{rad} B$	$x \notin \operatorname{rad} B$
84_{5}	Banach algebra	Banach algebra map
169^{14}	space L_p	space L^{∞}
$174_{9,\,8}$	The references entered as [2], [3] under J. H. Williamson should be [2], [3] under J. Wermer.	

W. Żelazko, *Banach Algebras*

Preface to the Polish edition

Banach algebras are Banach spaces equipped with a continuous binary operation of multiplication. Numerous spaces considered in functional analysis are also algebras, e.g. the space $C(0, 1)$ with pointwise multiplication of functions, or the space l_1 with convolution multiplication of sequences. Theorems of the general theory of Banach algebras, applied to those spaces, yield several classical results of analysis, e.g. the Wiener theorem and the Wiener–Levy theorem on trigonometric series (see Exercises 9.4.b and 16.6.a), or theorems on the spectral theory of operators. The foundations of the theory of Banach algebras are due to Gelfand. It was his astonishingly simple proof of the Wiener theorem that first turned the attention of mathematicians to the new theory. Certain specific algebras had been studied before, e.g. algebras of endomorphisms of Banach spaces, or weak-closed subalgebras of the algebra of endomorphisms of Hilbert spaces (the so-called von Neumann algebras or W^x algebras); also certain particular results had been obtained earlier. But the first theorem of the general theory of Banach algebras was the theorem on the three possible forms of normal fields, announced by Mazur in 1938. This result, now known as the Gelfand–Mazur theorem, is the starting point of Gelfand's entire theory of Banach algebras. We give Mazur's original proof in this book; it is its first publication.

The reader of this book is supposed to have some knowledge of functional analysis, algebra, topology, analytic functions and measure theory. The book is intelligible to first year university students, though certain sections are based on material which grew beyond the usual program of university mathematics, e.g. Chapter III makes use of the theory of analytic functions of several variables. However, the necessary notions and facts are always given explicity and provided with references.

The book consists of five chapters. The first two give an outline of the general theory of Banach algebras. Chapter III deals with analytic functions in Banach algebras, Chapter IV is devoted to involution algebras and Chapter V to function algebras. Some parts of the book may be omitted at the first reading without impairing the comprehension of the general theory. This concerns, in particular, Sections 4, 5 (from Theorem 5.5 on), 6, 7, 11, 14, 15 (the last two only at the initial stage of the reading), Chapter III except Section 16, Chapter IV except Sections 23, 24 and the whole of Chapter V. Most of the theorems and lemmas and also some of the definitions are followed by exercises. They not only provide an illustration for the preceding material, but also contain results of which use is made in subsequent considerations. Therefore they should be done or, at least, read through.

Chapters I–IV developed from lectures given by the author at the University of Warsaw in 1965/66. The reader will certainly notice the almost complete lack of information on group algebras and also the rather small amount of material concerning the algebras of operators. These omissions are due to shortage of space and may be justified by the existence of extensive literature concerning the above topics. However, it should be stressed that both abstract harmonic analysis and the theory of operators are the main source of inspiration for the general theory of Banach algebras, and the principal domain of its applications.

I wish to express my gratitude to Professor Stanisław Hartman, who read the text and introduced many essential corrections. Any inaccuracies that may still be found in the book can only be explained by my apparent failure to follow his suggestions precisely enough. I also wish to thank Professor Stanisław Mazur for his kind permission to publish his theorem together with its original proof.

W. Żelazko

Preface to the English edition

This is the second edition of the book, prepared four years after the writing of the first Polish edition. Since during that time several new results have been obtained and some other ones have proved to be of interest, I have added some new material. Thus, in § 6 the results on factorization (6.4–6.5) and in § 14 those on permanently singular elements (14.5–14.8) are added. The whole of § 18 is new and so are the results on the maximum modulus principle in § 19 (19.8–19.11). Single theorems are added in § 25 (25.9) and in § 29 (29.4). The whole of § 33 is new. There are also some minor improvements and rearragments of the material. I have omitted the material concerning von Neumann algebras since there are several books in English or French concerning this topic, e.g. Diximier [1], Sakai [1], or Schwartz [1], while there were none in Polish.

The main change, however, consists in adding, at the end of each chapter, comments on the possibility of generalizing the results into topological algebras, more general than Banach algebras. I have considered mainly locally bounded algebras and locally convex algebras, in particular multiplicatively convex and locally convex algebras. Information is also given on open questions concerning these generalizations. The comments reflect, of course, the author's own knowledge and taste (just as the rest of the book does): they are abundant with regard to the first three chapters and rather meagre with regard to the remaining ones. There is an appendix at the end of the book introducing the terminology used in the comments. We adopt the terminology of Banach [1] co-ordinated with the terminology of Dunford and Schwartz [1], so that, for example, an F-space or a Fréchet space means here a completely metrisable linear topological space, no local convexity being assumed. A locally convex F-space is called a B_0-space according to the terminology adopted in Mazur and Orlicz, [1] and [2], and used by Banach himself.

Finally I would like to express my gratitude to Marcin E. Kuczma, who translated the book and made numerous valuable suggestions concerning its improvement.

Warsaw, February 1972 *W. Żelazko*

CHAPTER I

General Information about Banach Algebras

§ 1. NOTATION, BASIC DEFINITIONS AND THEOREMS

The aim of this section is to establish the notation and terminology and to recall several notions and theorems necessary for the comprehension of the sequel. We assume that they are essentially known to the reader; however, some of them will be provided with references to the literature. This concerns particularly certain notions and theorems which are not generally known and which will find an application in more difficult parts of the book.

We now introduce notions and theorems of set theory, topology, algebra, functional analysis, and measure theory.

1.1. AGLEBRA OF SETS, MAPPINGS. We use the following symbols for the set theoretic operations: $A \cup B$ denotes the union, $A \cap B$ the intersection, $A \setminus B$ the difference of sets A and B (the complement of a set $A \subset X$ will be written as $X \setminus A$). The symbols $\bigcup\limits_{\alpha} A_\alpha$, $\bigcup\limits_{\alpha \in \mathfrak{A}} A$, $\bigcup\limits_{A \in \mathfrak{A}} A$ or $\bigcup \mathfrak{A}$ $(= \bigcup\limits_{A \in \mathfrak{A}} A)$ will stand for infinite unions; similarly for intersections. The class of all subsets of a set X will be denoted by 2^X. The Cartesian product of two sets A and B will be denoted by $A \times B$ and an infinite product by $\prod\limits_{\alpha \in \mathfrak{A}} A_\alpha$. The latter will also be written as $A^{\mathfrak{A}}$ whenever all A_α's are identical with some set A. The subset of all elements x of a set X which have some property W (i.e., for which $W(x)$ holds) will be denoted by $\{x \in X: W(x)\}$. The symbol $\{x_1, x_2, \ldots, x_n\}$ denotes the set consisting of elements $x_1, x_2, \ldots, x_n$; $\emptyset$ is the empty set.

The fact that a mapping φ is defined on a set A and takes values in a set B will be written as $\varphi: A \to B$; the mapping itself will be denoted by φ or by the symbol $x \to \varphi(x)$, whereas the symbol $\varphi(x)$ will be reserved for the value of φ at a point x. The characteristic function χ_A of a subset $A \subset X$ is defined by the equality

$$\chi_A(x) = \begin{cases} 1 & \text{for} \quad x \in A, \\ 0 & \text{for} \quad x \in X \setminus A. \end{cases}$$

In particular, if X is the set N of all positive integers or the set of all integers (which will not be denoted by a specific symbol), then the characteristic function of a set $\{n\}$, consisting of the single point n, will be denoted by the *Kronecker symbol* $k \to \delta_{nk}$. The symbol B^A will be employed to denote the family of all mappings of a set A into a set B. This notation coincides with the former symbols $A^{\mathfrak{A}}$ and 2^X (the latter being jusified by an identification of subsets of X with their characteristic functions). If $x \in X = \prod_{\alpha \in \mathfrak{A}} X_\alpha$, we write $x = (x_\alpha)_{\alpha \in \mathfrak{A}}$ and the element $x_\alpha \in X_\alpha$ is called the *α-th coordinate of* x. The mapping $p_\alpha: X \to X_\alpha$ given by $x \to x_\alpha$ is called the *projection of* X *onto the axis* X_α. Let $\boldsymbol{R}$ denote the real line and $\boldsymbol{C}$ the complex plane. Mappings $\varphi: X \to \boldsymbol{R}$ and $\varphi: X \to \boldsymbol{C}$ will be called *real-* resp. *complex-valued functions* defined on X.

1.2. ORDERED SETS. A set X of elements $x, y, z, \ldots$ will be called an *ordered set* if a relation $x \prec y$ is defined for certain pairs of its elements such that the following conditions are fulfilled:

(i) $x \prec x$,

(ii) if $x \prec y$ and $y \prec z$, then $x \prec z$,

(iii) if $x \prec y$ and $y \prec x$, then $x = y$.

An ordered set is called a *linearly ordered set* if for any pair of its elements $x, y \in X$ either $x \prec y$ or $y \prec x$ holds. A linearly ordered subset of an ordered set X is also called a *chain* in X.

An element x_0 of an ordered set X is called the *least upper bound of a subset* $A \subset X$ if $y \prec x_0$ for all $y \in A$ and if the condition $z \succ y$ for all $y \in A$ implies $z \succ x_0$.

An element $x_0 \in X$ is called a *maximal element* if the condition $y \succ x_0$ implies $y = x_0$. A *minimal element* is defined similarly.

An ordered set $\mathfrak{A}$ is called a *directed set* whenever for any two elements $\alpha, \beta \in \mathfrak{A}$ an element $\gamma \in \mathfrak{A}$ exists such that $\gamma \succ \alpha$ and $\gamma \succ \beta$.

A linearly ordered set is said to be *well ordered* if every non-void subset has a minimal element. In a well ordered set each element has a successor (the following element) but need not have a predecessor.

1.3. THE AXIOM OF CHOICE. This axiom will be often used throughout the book, usually in the following alternative formulation, called the *Kuratowski–Zorn lemma*:

If X is an ordered set and every chain $A \subset X$ has a least upper bound in X, then there exist maximal elements in X. More precisely, *for any chain $A \subset X$ there exists a maximal element $x_A \in X$ such that $x_A \succ y$ for all $y \in A$.*

The axiom of choice is used in the proof of Zermelo's theorem stating that every set admits a well ordering. This theorem is usually employed jointly with the following *transfinite induction principle*:

If X is a well ordered set whose least element has a property W and if the fact that all elements preceding an element x_0 have property W implies that x_0 itself enjoys this property, then $W(x)$ holds for all $x \in X$.

Proofs involving transfinite induction usually begin with the words: "Arrange all the elements of X into a transfinite sequence...", a well ordering being meant thereby. Yet the Kuratowski–Zorn lemma nearly always is a more elegant and effective tool.

1.4. TOPOLOGICAL SPACES, NETS. By a *topological space* we mean a pair (X, τ), where τ is the family of all open subsets of X, satisfying the well-known axioms. The *interior* of a subset A of a topological space X will be denoted by $\operatorname{int} A$ and the *closure* of A by $\bar{A}$. The *boundary* of A is the set $\partial A = \bar{A} \cap \overline{X \setminus A}$. A *neighbourhood* of a point $x \in X$ is any set $U \subset X$ such that $x \in \operatorname{int} U$. A *base of neighbourhoods* of a point $x \in X$ may be regarded as a directed set (1.2) if the ordering $\beta \succ \alpha$ is defined by $\beta \subset \alpha$ for α, β neighbourhoods of x. Choosing a point x_α in each neighbourhood $\alpha \in \mathfrak{A}$, a neighbourhood base of x, we obtain a net of points $(x_\alpha)_{\alpha \in \mathfrak{A}}$. This net converges to x in the topology τ, or τ-converges to x, in symbols $x_\alpha \to x$. In general, a net is any system of points $(x_\alpha)_{\alpha \in \mathfrak{A}}$, indexed by a directed set $\mathfrak{A}$. Such a net is said *to converge to a point* x if for any neighbourhood U of x there exists an index $\alpha_U \in \mathfrak{A}$ such that $x_\alpha \in U$ for $\alpha \succ \alpha_U$. An element x is a *cluster point* of a set $A \subset X$ (i.e. $x \in \bar{A}$) if and only if there exists a net $(x_\alpha) \subset A$ which is τ-convergent to x.

1.5. COMPARISON AND EQUIVALENCE OF TOPOLOGIES. Let τ_1 and τ_2 be topologies defined in the same set X. The topology τ_1 is said to be *weaker* than τ_2, in symbols $\tau_1 \leqslant \tau_2$, if $\tau_1 \subset \tau_2$. Two topologies are *equivalent* if they consist of the same open sets, or, which is the same, if the classes of closed sets coincide. In terms of neighbourhood bases (not defined here) the topology defined by a neighbourhood base $\mathfrak{B}_1$ is weaker than that defined by a base $\mathfrak{B}_2$ if for any point $x \in X$ and any $U \in \mathfrak{B}_1$, a neighbourhood of x, there exists a neighbourhood U' of x with $U' \in \mathfrak{B}_2$, $U' \subset U$. Relations between topologies may also be characterized in terms of continuous mappings. If τ_1 and τ_2 are topologies in a set X, then $\tau_1 \leqslant \tau_2$ holds if and only if for any topological space Y every τ_1-continuous function $\varphi: X \to Y$ is τ_2-continuous. (A mapping $\varphi: X \to Y$ is continuous if and only if the counter-image $\varphi^{-1}(U)$ of any open set $U \subset Y$ is open in X.)

1.6. THE WEAK TOPOLOGY INDUCED BY A FAMILY OF MAPPINGS. Let a family of topological spaces X_α, $\alpha \in \mathfrak{A}$, be given, and let a family of mappings of a set X into X_α be defined for each $\alpha \in \mathfrak{A}$. Then there exists the weakest topology in X in which all mappings in question are continuous. The base for this topology is formed by all finite intersections of counter-images (under these mappings) of open sets in X_α's.

1.7. HAUSDORFF SPACES; REGULAR SPACES; NORMAL SPACES. A topological space X is called a *Hausdorff space* if every two distinct points of X have disjoint neighbourhoods. A topological space X is called *regular* if for any point

$x \in X$ and any closed set $F \subset X$ with $x \notin F$ there exist disjoint open sets U_1 and U_2 such that $x \in U_1$, $F \subset U_2$. A space X is called *normal* if for every two disjoint closed sets $F_1, F_2 \subset X$ there exist disjoint open sets $U_1, U_2 \subset X$ containing F_1 and F_2 respectively. The following *Urysohn lemma* holds:

If X is a normal space and F_1, F_2 are disjoint closed non-void subsets of X, then there exists a continuous real-valued function f defined on X such that $f(p) = 1$ for $p \in F_1$, $f(p) = 0$ for $p \in F_2$, and $0 \leqslant f(p) \leqslant 1$ for all $p \in X$.

1.8. COMPACT SPACES. Each of the three following equivalent properties may be regarded as the definition of a *compact space*:

(i) Every open cover of X (= covering by open sets) contains a finite subcover.

(ii) Every family (F_α) of closed subsets of X whose intersection is void contains a finite subfamily whose intersection is void.

(iii) If (F_α) is any centered family of closed non-void subsets of X, then the intersection $\bigcap F_\alpha$ is non-void.

Compact spaces enjoy some additional nice properties:

Every compact Hausdorff space is normal.

The image of a compact space under a continuous mapping is a compact space.

If τ_1 and τ_2 are Hausdorff topologies in a set X such that the space (X, τ_1) is compact and $\tau_2 \leqslant \tau_1$, then $\tau_1 = \tau_2$. It hence follows that if φ is a continuous one-to-one mapping of a compact Hausdorff space onto another Hausdorff space, then φ is a homeomorphism.

We shall need in the sequel the following theorem on products of compact spaces.

1.9. PRODUCT TOPOLOGY; THE TICHONOV THEOREM. If X_α, $\alpha \in \mathfrak{A}$, is a family of topological spaces, then the product $X = \prod_{\alpha \in \mathfrak{A}} X_\alpha$ can be given a topology (called the *product topology* or the *Tichonov topology*) defined as the weakest topology in which all projections $p_\alpha \colon X \to X_\alpha$ are continuous (see 1.1 and 1.6). The local base at a point $x_0 \in X$ consists of neighbourhoods of the form

$$V = \{x \in X \colon p_{\alpha_i}(x) \in V_{\alpha_i},\ i = 1, \ldots, n\},$$

where $\alpha_1, \ldots, \alpha_n \in \mathfrak{A}$ and V_{α_i} is a neighbourhood of $p_{\alpha_i}(x_0)$ in X_{α_i}. The following *Tichonov theorem* is equivalent to the axiom of choice:

The product of any family of compact spaces is a compact space.

1.10. LOCALLY COMPACT SPACES; ONE-POINT COMPACTIFICATION. A topological space X is called *locally compact* if every point of X has a compact neighbourhood. An example of a locally compact space is any compact Hausdorff space with one point removed. Conversely, every locally compact and non-compact Hausdorff space may be homeomorphically embedded in a compact space by adjoining a single point p_∞ called "the point at infinity", i.e. by putting $X_1 = X \cup \cup \{p_\infty\}$ and defining the local base at p_∞ as the family of sets $\{(X \setminus Z) \cup \{p_\infty\}\}$ with Z ranging over all compact subsets of X. The space X is then said to be *compactified* by the point at infinity. We say that a continuous (real- or complex-valued) function f defined on a locally compact space X *vanishes at infinity* if for every $\varepsilon > 0$ there exists a compact subset $Z \subset X$ such that $|f(p)| < \varepsilon$ for $p \in X \setminus Z$. Such functions extend to continuous functions defined on the compactified space $X_1 = X \cup \{p_\infty\}$ whose value at the point p_∞ is zero. The set of those functions will be denoted by $C_0(X) = C_0(X_1 \setminus \{p_\infty\})$.

1.11. THE WEIERSTRASS–STONE THEOREM. *Let X be a compact Hausdorff space and let A be a family of continuous complex-valued functions defined on X such that*

(i) *For any* $p_1, p_2 \in X$, $p_1 \neq p_2$, *there exists a function* $f \in A$ *with* $f(p_1) \neq f(p_2)$ (i.e. *A separates the points of X*, or *A is a separating family*).

(ii) *For every* $f \in A$ *the conjugate function* $\bar{f}$ *also is in A* (the *conjugate function* $\bar{f}$ is the function whose value at any point $p \in X$ is equal $\overline{f(p)}$, the complex conjugate of $f(p)$).

Then the family of functions which can be uniformly approximated by complex polynomials of the elements of A (i.e. by complex linear combinations of finite products of elements of A) *contains either all continuous functions on X or all functions in* $C_0(X \setminus \{p_\infty\})$ *for some* $p_\infty \in X$.

In other words, *the subalgebra of* $C(X)$ *spanned by A is uniformly dense either in* $C(X)$ (this happens, in particular, when the family A contains a constant function) *or in some subspace of the form* $C_0(X \setminus \{p_\infty\})$ (this happens when all functions in A vanish at some point p_∞).

Let us make here the convention that, for a compact space X, the symbol $C(X)$ will always denote the space (the algebra) of all continuous complex-valued functions defined on X with the topology of uniform convergence (the sup norm). The space of all continuous real-valued functions defined on X will be denoted by $C_R(X)$. A real analogue of the Weierstrass–Stone Theorem may be formulated as follows:

If A is a separating family of continuous real-valued functions defined on a compact Hausdorff space X, then either the smallest closed subalgebra of $C_R(X)$ *containing the family A is the whole of* $C_R(X)$ (if, e.g., A contains constant functions) *or there exists a point* $p \in X$ *such that this subalgebra is precisely the set* $\{f \in C_R(X): f(p) = 0\}$ (Dunford and Schwartz [1], Chapter IV, Theorems 16 and 17).

1.12. VECTOR SPACES, ALGEBRAS, IDEALS. We are not defining here the concepts of a vector space, a linear map and linear independence; we just recall that an endomorphism of a vector space X is any linear mapping $f\colon X \to X$. An *algebra* is any vector space A equipped with an associative binary operation of multiplication satisfying the condition $(\alpha x)(\beta y) = (\alpha\beta)(xy)$ for any elements $x, y \in A$ and any scalars α, β. We shall restrict attention to spaces and algebras over the field of the reals or of the complexes; we shall refer to them as *real*, resp. *complex algebras* (*spaces*). An example of an algebra is the space of all endomorphisms of some vector space X with multiplication defined as composition of operators. A subalgebra I of an algebra A is called a *right ideal* if $IA \subset A$, and a *left ideal* if $AI \subset A$. Here we adopt the notation:

$$PQ = \{xy \in A\colon x \in P, y \in Q\}, \qquad \alpha P+\beta Q = \{\alpha x+\beta y\colon x \in P, y \in Q\}$$

for $P, Q \subset A$. An ideal I is said to be *proper* if $\{0\} \neq I \neq A$. An ideal which is both left and right is called a *two-sided ideal.* In a commutative algebra every ideal is two-sided.

1.13. QUOTIENT ALGEBRAS. Let I be a proper two-sided ideal of an algebra A. Subsets of A of the form $x+I$ with $x \in A$ are called *cosets modulo I.* Two distinct cosets always are disjoint; their sum also is a coset. From the fact that I is a two-sided ideal it follows that the product $(x+I)(y+I)$ of any two cosets is always contained in some coset which will be called the *product of cosets* $(x+I)$ *and* $(y+I)$. Finally, a scalar multiple of a coset is itself a coset. Thus the set of all cosets modulo I has the structure of an algebra. This algebra will be denoted by the symbol A/I and will be called the *quotient algebra of A modulo I.* The mapping $\varphi\colon A \to A/I$ given by $x \to x+I$ will be called the *natural homomorphism of A onto A/I* (we are not defining here the well-known concepts of a homomorphism or an isomorphism of groups or algebras).

A proper ideal $M \subset A$ is called a *maximal ideal* if it is not properly contained in any other proper ideal. In a commutative algebra with unit every proper ideal is contained in some maximal ideal; the same holds for non-commutative algebras, with the observation that left and right ideals should be treated separately (the corresponding maximal ideals also are left or right). The proof of those facts is an immediate consequence of the Kuratowski–Zorn lemma. A quotient algebra A/M of any commutative algebra A with unit modulo a maximal ideal M is a field, i.e. every non-zero element of A/M has an inverse.

We now recall some notions and facts from functional analysis. The concepts of a Banach space, a linear functional, the conjugate (dual) space, a convex set, a locally convex space, a topological vector space are considered as familiar and are not defined here. We use the convention that the expression "linear operators" or "functionals" always refers to continuous (bounded) operators and functionals; otherwise (if no continuity is assumed) we shall speak of *homogeneous and additive* operators or functionals.

1.14. DECOMPOSITION OF A BANACH SPACE INTO A DIRECT SUM. Let X be a Banach space, X_1, X_2 subspaces of X (a *subspace* is always supposed to be closed;

otherwise we shall speak of a *linear subset*). If X is the direct sum $X = X_1 \oplus X_2$, i.e. every element $x \in X$ can be written uniquely in the form $x = x_1 + x_2$ with $x_i \in X_i$, $i = 1, 2$, then the projections $p_i : X \to X_i$, defined by $p_i(x) = x_i$, $i = 1, 2$, are linear operators. This fact will be employed in some proofs of continuity of operators by means of a construction of a suitable decomposition.

1.15. Linear functionals in locally convex spaces. Let X be a locally convex real vector space, let K be a closed convex subset of X and let $p \in X \setminus K$. Then there exists a linear functional f such that $f(p) < \inf_{x \in K} f(x)$ or, which is the same, there exists a (closed) hyperplane (a translate of a subspace of codimension 1) which separates p from K (here e.g. the set $\{x \in X: f(x) = \frac{1}{2}[f(p) + \inf_{x \in K} f(x)]\}$ is such a hyperplane). If, further, the set K contains no subspace of X, then the set of such functionals f which separate K from points lying off K is a *total family of functionals*, i.e. such that the condition $f(x) = 0$ for all those functionals f implies $x = 0$.

A homogeneous and additive functional f on X is continuous if and only if the null-set $\{x \in X: f(x) = 0\}$ is closed in X.

1.16. The Banach closed graph theorem. *Let T: $X \to Y$ be a homogeneous additive operator mapping a Banach space X into a Banach space Y. If the graph $\{(x, Tx) \in X \times Y: x \in X\}$ is a closed subset of the product $X \times Y$, then T is a linear operator.*

1.17. The Banach inverse mapping theorem. *If X, Y are Banach spaces and if a linear operator T: $X \to Y$ is one-to-one and onto, then the inverse mapping T^{-1} also is a linear operator.*

It hence follows, in particular, that if $\|\cdot\|_1$ and $\|\cdot\|_2$ are two comparable norms in X (i.e. such that e.g. $\|x\|_1 \leqslant C\|x\|_2$, $x \in X$) and if X is a Banach space both in the norm $\|\cdot\|_1$ and in $\|\cdot\|_2$, then the two norms are in fact equivalent (i.e. another constant $C_1 > 0$ exists such that $C_1\|x\|_2 \leqslant \|x\|_1 \leqslant C\|x\|_2$, $x \in X$).

1.18. The weak topology. Let X be a Banach space, X' its dual. The weak topology $\sigma(X, X')$ is the weakest topology for X in which all functionals $f \in X'$ are continuous. This topology is defined by a local base of neighbourhoods of zero, consisting of neighbourhoods of the form

$$V = \{x \in X: |f_i(x)| < \varepsilon,\ i = 1, \ldots, n\},$$

where $f_1, \ldots, f_n \in X'$, $\varepsilon > 0$. The space $[X, \sigma(X, X')]$ is a locally convex space. Similarly, the weak topology $\sigma(X', X)$ of X' is defined by the local base formed by the neighbourhoods

$$V = \{f \in X': |f(x_i)| < \varepsilon,\ i = 1, \ldots, n\},$$

$x_i \in X$, $\varepsilon > 0$. This topology is to be distinguished from the weak topology $\sigma(X', X'')$.

It can be proved that the unit ball $K = \{f \in X' \colon \|f\| \leqslant 1\}$ is compact in the topology $\sigma(X', X)$.

1.19. THE KREIN–MILMAN THEOREM (cf. Dunford and Schwartz [1], Chapter V, Theorem 4). Let X be a locally convex space and let Z be a closed convex subset of X. A point $x \in Z$ is called an *extremal point of* Z if it is not an interior point of any segment contained in Z (i.e. $x \neq tx_0 + (1-t)x_1$ for all $x_0, x_1 \in Z$, $0 < t < 1$); equivalently, a point $x \in Z$ is extremal if and only if the equality $x = \frac{1}{2}(x_0 + x_1)$, $x_0, x_1 \in Z$, implies $x_0 = x_1 = x$. The *Krein–Milman theorem* asserts that:

If Z is a closed convex subset of a locally convex space X, then Z has extremal points and coincides with the closed convex hull of its extremal points (the *closed convex hull* of a set is the smallest closed and convex set containing the set in question).

This theorem will be applied chiefly to the space $[X', \sigma(X', X)]$, where X is some Banach space. Compact subsets of this space are just those sets which are weak-closed and norm-bounded.

We conclude this introductory section with two theorems on measures and measurable functions which will be needed in the sequel:

1.20. *If φ is a measurable real- or complex-valued function of a real variable t such that $\varphi(x+y) = \varphi(x)+\varphi(y)$ almost everywhere (with respect to the planar Lebesgue measure), then $\varphi(t) = Ct$.*

1.21. *If f is a measurable function such that $\int fg d\mu = 0$ for every absolutely summable function g, then $f = 0$ almost everywhere.*

Other necessary concepts and theorems will be formulated successively in the text. Some facts and notions concerning comments to chapters will be given in the Appendix.

§ 2. TOPOLOGICAL ALGEBRAS AND BANACH ALGEBRAS

2.1. DEFINITION. A *topological algebra* over the field K of the reals or of the complexes is a topological vector space A over K provided with an associative multiplication continuous with respect to the system of both variables (i.e. jointly continuous).

A topological algebra has thus a double-featured structure: the structure of a topological vector space and that of an algebra over K. They are connected with each other by the condition of continuity of the algebraic operations with respect to the topology. Topological algebras can be classified as the underlying topological vector spaces. So, e.g., we mean by metric algebras those topological algebras which, regarded as topological vector spaces, are metric topological vector spaces. Similarly we talk about locally convex algebras, locally bounded algebras and so on. According to this convention we give the following definition:

2.2. DEFINITION. A *Banach algebra* is a topological algebra which, regarded as a topological vector space, is a Banach space. Such an algebra will also be called an *algebra of type* **B** or a **B**-*algebra*, in symbols $A \in \mathbf{B}$.

2.3. DEFINITION. A homomorphism $\varphi: A_1 \to A_2$ between two given topological algebras A_1 and A_2 is called a *topological homomorphism* provided φ is continuous; an isomorphism $\psi: A_1 \to A_2$ is called a *topological isomorphism* provided ψ is a homeomorphism between A_1 and $\psi(A_1)$.

Thus a one-to-one topological homomorphism need not be a topological isomorphism (if the inverse is not continuous). In the sequel homomorphisms and isomorphisms between topological algebras will always denote topological ones. Otherwise, if no continuity is assumed, we shall speak of algebraic homomorphisms and isomorphisms.

We emphasize here the fact that only those properties of topological algebras will be dealt with, which are preserved under isomorphisms. And thus two isomorphic Banach algebras will be viewed upon as two realizations of the same algebra and a Banach algebra with one norm replaced by an equivalent one will be regarded as unchanged.

If a Banach algebra A has no unit, then it can be embedded in a Banach algebra A_1 with unit. In fact, setting $A_1 = A \oplus \{\lambda e\}$ and $||(x, \lambda)|| = ||x|| + |\lambda|$ we obtain a Banach algebra: it is a complete normed space as a direct sum of two Banach spaces. The verification of the continuity of multiplication will be left to the reader. Clearly $\tilde{A} = \{(x, 0): x \in A\}$ is a subalgebra (a two-sided ideal, in fact) of A_1, isomorphically isometric to A.

We now prove a theorem essential to our further considerations.

2.4. THEOREM. *Let A be a Banach algebra. Then there exists a Banach space X such that A is isomorphic to a closed subalgebra of B(X), the algebra of all bounded operators $X \to X$ with the operator-norm* $||T|| = \sup\limits_{||x||=1} ||Tx||$.

PROOF. Let $X = A$ if A has a unit, $X = A \oplus \{\lambda e\}$ otherwise. Thus X is a Banach algebra with a unit e and A is a (proper or not) subalgebra of X. Let φ denote the map $A \to B(X)$ given by

$$\varphi(x) = T_x,$$

where $T_x \in B(X)$ is the operator of the left-hand multiplication by the element $x \in A \subset X$: $T_x y = xy$. As is easy to see, $T_{\lambda x+\mu y} = \lambda T_x + \mu T_y$ and $T_{xy} = T_x T_y$ for $x, y \in A$, $\lambda, \mu \in K$, where K is the field of scalars (real or complex). Further, if $T_x = 0$ in $B(X)$, i.e. if $T_x y = 0$ for each $y \in X$, then $x = 0$, which can be verified by putting $y = e$. The map φ is thus an algebraic isomorphism. It remains to verify that φ is a homeomorphism or, equivalently, that the norm in A given by

(2.4.1) $$|||x||| = ||T_x||$$

is equivalent to the original norm in A. We have

$$|||x||| = ||T_x|| = \sup_{||y|| \leqslant 1} ||xy|| \geqslant \left\| \frac{xe}{||e||} \right\| = \frac{||x||}{||e||},$$

so

(2.4.2) $$||x|| \leqslant ||e||\, |||x|||.$$

The norms $|||\cdot|||$ and $||\cdot||$ are thus comparable and the equivalence will be proved if we show that the algebra A is complete in the norm $|||\cdot|||$ (see 1.17) or, which is the same, that the subalgebra $\tilde{A} = \{T_x \in B(X)\colon\ x \in A\}$ is closed in $B(X)$. Let a sequence of operators T_{x_n} converge in $B(X)$ to an operator T. For any $y, z \in X$ we have $T_{x_n}(yz) = T_{x_n}(y)z$. Since the convergence of a sequence of operators implies the convergence of the sequence of values at each point and since the multiplication is continuous, by passing to the limit, we obtain the relation

$$T(yz) = T(y)z$$

for any $y, z \in X$. Putting $y = e$ we obtain $T(z) = T(e)z$. On the other hand, $T(e) = \lim T_{x_n}(e) = \lim x_n = x_0$; the last limit exists in A by (2.4.2) and the fact that the sequence x_n converges in the norm $|||\cdot|||$. We thus have $T(x) = T_{x_0}(x)$ for all $x \in X$, i.e. $T \in \tilde{A}$, which proves the equivalence of the two norms.

2.5. COROLLARY. *In every Banach algebra A there exists a norm equivalent to the original norm in A and satisfying the condition*

(2.5.1) $$||xy|| \leqslant ||x||\,||y||,$$

and

(2.5.2) $$||e|| = 1$$

whenever A has a unit e.

2.6. REMARK. In the proof of Theorem 2.4 we made no use of the fact that the multiplication in a Banach algebra is jointly continuous; we employed only the fact that it is a continuous linear map when one of the variables is fixed. Therefore several authors define a Banach algebra as a Banach space provided with a multiplication continuous with respect to each variable (separately continuous). It follows from the proof of Theorem 2.4 that both definitions are equivalent.

2.7. REMARK. In the proof of Theorem 2.4 X could not be taken simply as A. For if A had no unit, then the map φ need not be an algebraic isomorphism. E.g., if we let $xy = 0$ for all $x, y \in X$, X a Banach space, we obtain a Banach algebra with the so-called *trivial multiplication*. Then clearly $T_x = 0$ for $x \in X$.

2.8. Remark. Numerous authors define a Banach algebra as a Banach space provided with an associative multiplication satisfying conditions (2.5.1) and (2.5.2). This method is disadvantageous in that these conditions are not invariant under topological isomorphisms; besides, it involves a fixed norm in an algebra (though several equivalent norms can satisfy these conditions). Such a definition suits better those considerations which are dealing with metric (norm-preserving) invariants of considered algebras.

2.9. Remark. In the literature on the theory of Banach algebras one usually calls a *normed algebra* a topological algebra which is a normed space and whose norm satisfies conditions (2.5.1) and (2.5.2). It can be shown that this definition is equivalent to that which defines a normed algebra as a topological algebra which is a normed space and whose multiplication is jointly continuous. Then there exists in A a *submultiplicative norm* (i.e. one satisfying (2.5.1)) equivalent to the original norm. In the Russian literature Banach algebras are traditionally called *normed rings* (*normirovannye kol'ca*). But also there this latter name is now abandonned and gets replaced by the generally used expression.

§ 3. EXAMPLES OF BANACH ALGEBRAS

We give no proofs in our examples. The reader may regard them as exercises.

3.1. Let Ω be a compact Hausdorff space. Let $C(\Omega)$ denote the space of all continuous complex-valued functions defined in Ω. It is a Banach space under the norm $\|x\| = \max_{t\in\Omega}|x(t)|$. Since clearly $\|xy\| \leqslant \|x\|\,\|y\|$, then $C(\Omega)$ is a Banach algebra with multiplication defined pointwise. It is a commutative algebra with unit.

3.2. Let Ω be a topological Hausdorff space. Let $C_B(\Omega)$ denote the Banach space of all continuous and bounded functions on Ω with the norm $\|x\| = \sup_{t\in\Omega}|x(t)|$. It is a Banach algebra under pointwise multiplication.

3.3. Let $L_1(-\infty, \infty)$ denote the Banach space of functions absolutely summable on the real line, with the norm

$$\|x\| = \int_{-\infty}^{\infty} |x(t)|\,dt.$$

The multiplication in $L_1(-\infty, \infty)$ is defined as the convolution

$$x * y = \int_{-\infty}^{\infty} x(t-\tau)y(\tau)\,d\tau. \tag{3.3.1}$$

The convolution is commutative and associative, moreover,

$$\|x * y\| = \int_{-\infty}^{\infty} \left| \int_{-\infty}^{\infty} x(t-\tau)\, y(\tau) \right| d\tau\, dt \leqslant \int_{-\infty}^{\infty} \int_{-\infty}^{\infty} |x(t-\tau)|\, |y(\tau)|\, d\tau\, dt$$

$$= \int_{-\infty}^{\infty} |y(\tau)| \int_{-\infty}^{\infty} |x(t-\tau)|\, dt\, d\tau = \|x\|\, \|y\|,$$

so $L_1(-\infty, \infty)$ is a Banach algebra. It can be verified that $L_1(-\infty, \infty)$ has no unit (see Exercise 6.3.a).

3.3.a. EXERCISE. **Determine if $L_1(0, \infty) = \{x \in L_1(-\infty, \infty): x(t) = 0$ for $t \leqslant 0\}$ is a subalgebra of $L_1(-\infty, \infty)$. Show that the set of elements of $L_1(0, \infty)$ vanishing in a neighbourhood (depending on the element) of the point $t = 0$ is an ideal in $L_1(0, \infty)$.**

3.4. Let $L_1(0, 1)$ denote the Banach space of complex-valued functions absolutely summable on the interval (0, 1), with the norm

$$\|x\| = \int_0^1 |x(t)|\, dt.$$

It is a commutative Banach algebra without unit if the multiplication is defined as the convolution

$$x * y(t) = \int_0^1 x(t-\tau)\, y(\tau)\, d\tau, \qquad 0 < t < 1,$$

since $\|x * y\| \leqslant \|x\|\, \|y\|$, as is easy to verify.

3.5. Let $l_1(-\infty, \infty)$ denote the Banach space of two-sided sequences of complex numbers $\alpha = \{\alpha_n\}_{n=-\infty}^{\infty}$ with the norm

$$\|\alpha\| = \sum_{n=-\infty}^{\infty} |\alpha_n|.$$

It is a commutative Banach algebra if the multiplication is defined as the convolution

$$(\alpha * \beta)_n = \sum_{k=-\infty}^{\infty} \alpha_{n-k}\, \beta_k,$$

i.e. $\alpha * \beta$ is the sequence whose n-th entry is equal to $(\alpha * \beta)_n$. It can be easily verified that $l_1(-\infty, \infty)$ is an algebra with unit. The unit element is the sequence $\{\delta_{0n}\}$,

$$\delta_{ij} = \begin{cases} 0 & \text{for} \quad i \neq j, \\ 1 & \text{for} \quad i = j. \end{cases}$$

3.5.a. EXERCISE. **Show that $l_1(0, \infty) = \{\alpha \in l_1(-\infty, \infty): \alpha_n = 0$ for $n < 0\}$ is a subalgebra of the algebra $l_1(-\infty, \infty)$.**

3.6. Let $\omega = \omega(t)$, $-\infty < t < \infty$, be a non-negative function satisfying the inequality

$$\omega(t_1+t_2) \leqslant \omega(t_1)\omega(t_2) \tag{3.6.1}$$

for all real t_1, t_2.

Denote by $L_1^{\langle\omega\rangle}(-\infty, \infty)$ the Banach space of functions absolutely summable with the weight ω, that is the set of all functions $x(t)$ such that

$$\|x\| = \int_{-\infty}^{\infty} |x(t)|\,\omega(t)\,dt < \infty.$$

$L_1^{\langle\omega\rangle}(-\infty, \infty)$ is a commutative Banach algebra with the convolution (3.3.1).

In particular, we can take as $\omega(t)$ the function

$$\omega(t) = \begin{cases} e^t & \text{for} \quad t \geqslant 0, \\ 1 & \text{for} \quad t < 0. \end{cases}$$

Similarly we define the algebra $L_1^{\langle\omega\rangle}(0, \infty)$, where ω is a function satisfying (3.6.1) for non-negative t, and the algebras $l_1^{\langle\omega_n\rangle}(-\infty, \infty)$ and $l_1^{\langle\omega_n\rangle}(0, \infty)$, where $\{\omega_n\}$ is a sequence of non-negative numbers satisfying the condition analogous to (3.6.1):

$$\omega_{n+k} \leqslant \omega_n \omega_k,$$

which should hold for all integers n, k in the case of the algebra $l_1^{\langle\omega_n\rangle}(-\infty, \infty)$ and for non-negative integers in the case of the algebra $l_1^{\langle\omega_n\rangle}(0, \infty)$.

3.7. Let $L_1(K)$ denote the Banach space of all absolutely summable complex-valued functions defined on the unit circle K of the complex plane, $K = \{z \in \boldsymbol{C}\colon |z| = 1\}$ (K is a group with respect to multiplication). It is easy to verify that $L_1(K)$ is a Banach algebra with the convolution multiplication

$$x * y(t) = \int_K x(t\tau^{-1})\,y(\tau)\,\mu(d\tau) \tag{3.7.1}$$

and the norm

$$\|x\| = \int_K |x(t)|\,\mu(dt),$$

where μ is the Lebesgue measure on K.

3.7.a. Exercise. Denote by A the algebra whose elements are complex-valued periodic functions (with period 2π) defined on the real line and such that

$$\|x\| = \int_0^{2\pi} |x(t)|\,dt < \infty.$$

Prove that A is a Banach algebra under the convolution

$$x * y(t) = \int_0^{2\pi} x(t-\tau)\,y(\tau)\,d\tau$$

and that the algebra A is isomorphically isometric with the algebra $L_1(K)$.

3.7.b. Exercise. Prove that the algebra $L_1(K)$ has no unit.

3.8. REMARK. The algebras from the Examples 3.3, 3.5, 3.7 are particular cases of the so-called *group algebras*. As is well known, for every locally compact group G there exists a Borel measure μ on G called the *Haar measure* (more precisely, the *left-invariant Haar measure*) satisfying the following conditions:

(3.8.1) $\mu(xE) = \mu(E)$ for every $x \in G$ and every measurable $E \subset G$,

(3.8.2) $\mu(U) > 0$ for every non-void open set $U \subset G$,

(3.8.3) $\mu(K) < \infty$ for every compact set $K \subset G$.

The above conditions determine the measure μ uniquely up to a positive factor.

The *group algebra* of the group G is the space $L_1(G)$ of all complex-valued functions absolutely summable on G with respect to the Haar measure, with the norm

$$\|x\| = \int_G |x(t)|\,\mu(dt)$$

and with the convolution multiplication defined by formula (3.7.1) with G in place of K (cf. Gelfand, Raikov, Shilov [1], and Naimark [1]).

3.9. Let $\mathscr{A}$ denote the algebra of all functions holomorphic in the open disc $K = \{z \in \boldsymbol{C}: |z| < 1\}$ and continuous on its closure. It is a Banach algebra under the norm

$$\|x\| = \max_{|t| \leqslant 1} |x(t)| = \max_{|t|=1} |x(t)|$$

and pointwise multiplication of functions. The algebra $\mathscr{A}$ may be regarded as a subalgebra of the algebra $C(\bar{K})$ as well as a subalgebra of $C(\Gamma)$, where $\Gamma = \{z \in \boldsymbol{C}: |z| = 1\}$.

3.10. Denote by H_∞ the algebra of all bounded holomorphic functions in the unit disc K with pointwise multiplication and the norm $\|x\| = \sup|x(t)|$. It is a Banach algebra, a superalgebra of $\mathscr{A}$.

3.11. Let X be a Banach space. Denote by $B(X)$ the algebra of all bounded linear operators $X \to X$ with multiplication defined as composition (this algebra was already considered in the proof of Theorem 2.4). It is a Banach algebra with the norm $\|T\| = \sup_{\|x\| \leqslant 1} \|Tx\|$. This is an algebra with unit and is non-commutative unless $\dim X = 1$. As shown in Theorem 2.4, every Banach algebra is a subalgebra of such an algebra. In particular, letting X be the complex Euclidean space $\boldsymbol{C}^n$ we obtain the algebra of all complex matrices $n \times n$.

§ 4. METHODS OF CONSTRUCTION OF BANACH ALGEBRAS

In this section we show how new Banach algebras can be made out of old ones.

4.1. MULTIPLICATION WITH A WEIGHT.

4.1.a. Exercise. Let A be a commutative Banach algebra; fix an element $a \in A$. Define a new multiplication in A by: $x \times_a y = axy$. Show that A is a Banach algebra with respect to the multiplication $\times_a$. Determine when an algebra A with multiplication $\times_a$ has a unit.

4.2. The Arens multiplication. Let A be a Banach algebra. Denote by A', A'' the first and the second conjugate space of A. Let $\langle f, x\rangle$, $f \in A'$, $x \in A$ be the functional in A' defined by: $\langle f, x\rangle(y) = f(xy)$, $y \in A$. Let $F \in A''$, $f \in A'$; we define $[F,f] \in A'$ by $[F,f](x) = F(\langle f, x\rangle)$. Finally, we define the product FG, $F, G \in A''$ by the formula $FG(f) = F([G,f])$. This product is called the *Arens product*.

4.2.a. Exercise. Prove that A'' is a Banach algebra if the multiplication is defined by the Arens product.

4.2.b. Exercise. Prove that the canonical embedding $\pi: A \to A''$ is an isometric isomorphism between the algebras A and $\pi(A) \subset A''$.

4.3. Direct sums of algebras. Given a finite number of Banach algebras $A_1, A_2, \ldots, A_n$, we can define their direct sum $A = A_1 \oplus A_2 \oplus \ldots \oplus A_n$. The elements of A are finite sequences $x = (x_1, x_2, \ldots, x_n)$, $x_i \in A_i$ with coordinatewise multiplication and with the norm $\|x\| = \max_{i \leqslant n} \|x_i\|$. It is easily seen that A is a Banach algebra; we call A the *direct sum of the algebras* $A_1, A_2, \ldots, A_n$.

4.3.a. Exercise. Let $A = A_1 \oplus A_2 \oplus \ldots \oplus A_n$. Show that $\tilde{A}_i = \{x \in A: x_k = 0 \text{ for } k \neq i\}$ is an ideal in A isometrically isomorphic to the algebra A_i.

4.3.b. Exercise. Prove that $C[0,1] \oplus C[0,1] = C([0,1] \cup [2,3])$, where the equality denotes an isometric isomorphism.

4.3.c. Exercise. Prove that the direct sum algebra A has a unit if and only if all A_i's have units.

4.4. Quotient algebras. Let A be a Banach algebra and let I be a proper two-sided ideal of A. Prove that the quotient algebra A/I is a Banach algebra if the norm is defined by $\|(x)\| = \inf_{y \in (x)} \|y\|$, where (x) denotes the coset in A/I containing x.

4.4.a. Exercise. Let $A = A_1 \oplus A_2$. Prove that $A_1 = A/A_2$ (A_2 means here $\tilde{A}_2$, see Exercise 4.3.a), where the equality denotes an isometric isomorphism.

4.4.b. Exercise. Let $A = C[0,1]$, $I = \{x \in A: x(0) = 0\}$. What is A/I?

4.5. Bounded direct sums. Let A_α, $\alpha \in \mathfrak{A}$, be any family of Banach algebras. Denote by $\sum_\alpha A_\alpha$ the subset of the Cartesian product of A_α's consisting of those elements $x = (x_\alpha)$ for which

$$\|x\| = \sup_\alpha \|x_\alpha\| < \infty. \tag{4.5.1}$$

4.5.a. Exercise. Prove that $A = \sum_\alpha A_\alpha$ is a Banach algebra under the norm (4.5.1) and coordinatewise multiplication.

4.6. COMPLEXIFICATION. Let A be a real Banach algebra with unit. Consider the set $\tilde{A}$ of all pairs (x, y), $x, y \in A$. The set A is a Banach space if addition and multiplication by scalars are defined coordinatewise and the norm in A is defined by the formula

$$||(x, y)|| = \sup_{\varphi}(||x\cos\varphi - y\sin\varphi|| + ||y\cos\varphi + x\sin\varphi||).$$

A is a Banach algebra under the multiplication defined by the formula

$$(x, y)(u, v) = (xu - yv, xv + yu).$$

It is a complex algebra if a complex scalar $\alpha = a+bi$ is identified with the pair $(ae, be) \in \tilde{A}$. The embedding $A \to \tilde{A}$ given by $x \to (x, 0)$ is a homeomorphic isomorphism in view of the inequalities $||x|| \leqslant ||(x, 0)|| \leqslant \sqrt{2}\,||x||$. In this way every real algebra can be isomorphically embedded in some complex algebra.

4.7. ALGEBRAS OF QUOTIENTS (Suciu [1]). Let A be a commutative Banach algebra with unit and a norm $||x||$. Write

$$S = \{x \in A\colon x \neq 0 \text{ and } ||xy|| = ||x||\,||y|| \text{ for every } y \in A\}.$$

Let $\tilde{A}$ denote the set of all pairs (x, y) (regarded as quotients x/y) such that $x \in A$, $y \in S$; we identify any two pairs (x, y), (x', y') such that $xy' = x'y$. $\tilde{A}$ is thus in fact the set of equivalence classes of pairs (x, y). $\tilde{A}$ is a normed algebra if the operations are defined as this is usually done with quotients, i.e. $(x, y)(x', y') = (xx', yy')$, $(x, y)+(x', y') = (xy'+x'y, yy')$, $\lambda(x, y) = (\lambda x, y)$, and the norm is defined by the formula

$$||(x, y)|| = \frac{||x||}{||y||}. \tag{4.7.1}$$

It can be easily verified that the operations and the norm are well defined, i.e. do not depend on the choice of representatives. The *algebra of quotients* of the algebra A is the completion $\bar{A}$ of $\tilde{A}$ in the norm (4.7.1). Clearly the algebra A is isomorphically isometric with a subalgebra of $\bar{A}$.

4.7.a. EXERCISE. Examine the set S in the case of the algebra $\mathscr{A}$ (Example 3.9) and find its algebra of quotients.

We omitted here an important method of forming new Banach algebras from the old ones, namely that of taking tensor products. We indicate here only few references concerning this topic, Gelbaum [1]–[3], Gil de Lamadrid [1]. The tensor product algebras form an important tool in harmonic analysis on locally compact groups (cf. Varopoulos [2], [3]). For tensor products of more general algebras cf. Mallios [1].

§ 5. THE MAZUR–GELFAND THEOREM

5.1. THEOREM. *The set G of all invertible elements of a Banach algebra A with unit is open.*

PROOF. We first prove that the elements of the open set $U = \{x \in A: \|e-x\| < 1\}$ are invertible (e is the unit of A). In fact, putting $y = e-x$ for $x \in U$ we have $\|y\| < \alpha < 1$ for some α, hence $\|y^n\| < \alpha^n$ and the series $\sum_{n=0}^{\infty} y^n$ converges absolutely in A. Write $z = \sum_{n=0}^{\infty} y^n$; we have

$$zx = z(e-y) = \sum_{n=0}^{\infty} y^n - \sum_{n=0}^{\infty} y^{n+1} = e.$$

Similarly $xz = e$, so $z = x^{-1}$. Now, let x be an arbitrary invertible element of A. The map $y \to xy$ is a homeomorphism of A onto itself, so it carries open sets onto open sets. It follows that xU is an open set in A. This set consists of invertible elements and contains x since $e \in U$. Thus any invertible element has a neighbourhood consisting entirely of invertible elements.

5.2. COROLLARY. *The closure of a proper ideal in a Banach algebra A with unit is a proper ideal.*

PROOF. It is easy to see that the closure of a proper ideal I is either a proper ideal or the whole algebra A. On the other hand a proper ideal contains no invertible elements, so $I \cap G = \emptyset$. Thus also $\bar{I} \cap G = \emptyset$ whence $\bar{I} \neq A$.

5.2.a. EXERCISE. Use the algebra $A = \{x \in C[0, 1]: x(0) = 0\}$ and the ideal $I = x_0 A$, where $x_0(t) = t$, to show that Corollary 5.2 does not hold for algebras without unit.

5.3. THEOREM. *Let A be a Banach algebra with unit; then the operation $x \to x^{-1}$ is continuous on the set G of all invertible elements of A.*

PROOF. Let $x_n \in G$, $n = 1, 2, \ldots$, and $\lim x_n = y \in G$. It must be shown that $\lim x_n^{-1} = y^{-1}$. Assume first that $y = e$; then there exists an integer N such that $\|e-x_n\| < \frac{1}{2}$ for $n > N$. We thus have the estimation

$$\|x_n^{-1}\| = \left\|\sum_{k=0}^{\infty} (e-x_n)^k\right\| \leqslant \sum_{k=0}^{\infty} \|e-x_n\|^k \leqslant \sum_{k=0}^{\infty} 1/2^k = 2$$

for $n > N$, so

$$\|x_n^{-1}-e\| = \|x_n^{-1}(e-x_n)\| \leqslant \|x_n^{-1}\| \, \|e-x_n\| \leqslant 2\|e-x_n\|$$

for $n > N$ whence $\lim x_n^{-1} = e$. Now let y be any invertible element of A. Since $y^{-1}x_n \in G$ and $\lim y^{-1}x_n = e$, then

$$\lim (y^{-1}x_n)^{-1} = \lim x_n^{-1} y = e \quad \text{and} \quad \lim x_n^{-1} = y^{-1}.$$

So far we have not fixed the field of scalars. Now we assume that it is the field of the complexes $\boldsymbol{C}$. We shall prove the famous Mazur–Gelfand theorem crucial for the whole theory of Banach algebras.

5.4. THEOREM. *Let A be a field (possibly non-commutative) of type* **B** *over* **C** *Then A is isomorphic to the field of complex numbers.*

PROOF. It suffices to show that each element $x \in A$ has the form $x = \lambda e$, where λ is a complex scalar. Suppose that this is not the case, i.e. that there exists an $x \neq \lambda e$ for all $\lambda \in C$. We thus have $x - \lambda e \neq 0$ for every $\lambda \in C$, so that the inverse $(x-\lambda e)^{-1}$ exists for $\lambda \in C$. Let f be a (continuous) linear functional on A such that $f(x^{-1}) \neq 0$ and write $\varphi(\lambda) = f([x-\lambda e]^{-1})$. We are going to show that $\varphi(\lambda)$ is an entire function. In fact,

$$\frac{\varphi(\lambda+h)-\varphi(\lambda)}{h} = f([x-\lambda e]^{-1}[x-(\lambda+h)e]^{-1});$$

let $h \to 0$; the limit of the right-hand side exists by Theorem 5.3. Moreover,

$$\varphi(\lambda) = \frac{1}{\lambda} f\left(\left[\frac{x}{\lambda} - e\right]^{-1}\right)$$

for $\lambda \neq 0$, and since

$$\lim_{|\lambda| \to \infty} \left(\frac{x}{\lambda} - e\right)^{-1} = -e,$$

so

$$\lim_{|\lambda| \to \infty} |\varphi(\lambda)| = 0.$$

In virtue of the Liouville theorem we have $\varphi(\lambda) \equiv 0$ which is impossible since $\varphi(0) = f(x^{-1}) \neq 0$. This is a contradiction to the supposition that $x \neq \lambda e$ for all $\lambda \in C$.

This theorem was first announced by Mazur [1] (1). The first published proof for complex scalars has been given by Gelfand in the paper [1] which also contains further basic facts of the theory of commutative Banach algebras. The above proof is due to Arens [2]. It is valid not only in Banach algebras. The only essential suppositions are those of the continuity of the map $x \to x^{-1}$ on the set $\{x \in A: x \neq 0\}$ and of the existence of a total family of linear functionals. This proof may be reduced also to the case when A possesses at least one non-trivial linear functional (cf. Żelazko [8]).

We now give Mazur's original proof for normed fields regarded as algebras over the field of the reals. This proof is based on the following *Frobenius theorem* (cf. e.g. Pontriagin [1]):

Every field, finite-dimensional over the field of real numbers is isomorphic either to the field of real numbers or to the field of complex numbers or to the division algebra of quaternions.

This theorem may be given a form more useful for our purposes:

---

(1) The original manuscript of this paper contained a complete proof. However, this made the paper too voluminous and the editors of Comptes Rendus required a more concise form. The only sensible way of a shortening was to leave out all the proofs, and the paper was finally so published.

If A is a linear field over the field of real numbers and every element x of A is algebraic, i.e. there exists a non-zero polynomial W with real coefficients such that $W(x) = 0$, then A is isomorphic to either of the three finite-dimensional fields.

5.5. THEOREM. *Let A be a Banach algebra over the field of real numbers and suppose that for every $x \in A$, $x \neq 0$ there exists an $x^{-1} \in A$ (i.e. A is a field). Then the algebra A is homeomorphically isomorphic either to the field of real numbers or to the field of complex numbers or to the division algebra of quaternions.*

PROOF. If A is a finite-dimensional field, then the assertion follows by the Frobenius theorem. If A is infinite-dimensional, then there exists a transcendental element in A, according to the second formulation of the Frobenius theorem. In this case A contains a field isomorphic to the field of all rational functions of a single variable with real coefficients. Now it suffices to show that no submultiplicative norm can be introduced in the field W of rational functions. The proof will be indirect. Suppose, on the contrary, that such a norm $\|\cdot\|$ exists in W. For any complex number $\zeta = \alpha+\beta i$, $\alpha, \beta \in R$, write

$$W_\zeta(t) = \mathrm{re}[(\zeta-t)^{-1}] = [(\alpha-t)^2+\beta^2]^{-1}(\alpha-t);$$

W_ζ is a rational function of the variable t with real coefficients. Clearly $W_1 \neq W_0$, so there exists a linear functional φ in W continuous with respect to the norm $\|\cdot\|$ and such that

$$\varphi(W_0) \neq \varphi(W_1). \tag{5.5.1}$$

Now put $f(\zeta) = \varphi(W_\zeta)$. $\zeta \to f(\zeta)$ is a continuous real-valued function defined in the complex plane C. We shall prove that the function f is harmonic in the closed plane $C \cup \{\infty\}$. By a well-known property of harmonic functions (the analogue of the Liouville theorem) it then follows that f is constant, which, by (5.5.1), yields the desired contradiction.

Let u_k, $v_k \in W$, $k = 1, 2, \ldots, n$. Setting $w_k(t) = u_k(t)+iv_k(t)$ we have $\mathrm{re}(w_1 w_2 \ldots w_n) \in W$. We shall need the following estimation:

$$\|\mathrm{re}(w_1 w_2 \ldots w_n)\| \leqslant (\|u_1\|+\|v_1\|)(\|u_2\|+\|v_2\|) \ldots (\|u_n\|+\|v_n\|). \tag{5.5.2}$$

The proof will be by induction on n. For $n = 1$ the inequality is obvious; suppose that it holds for some $n = k$. We have, by the induction assumption,

$$\begin{aligned}\|\mathrm{re}(w_1 w_2 \ldots w_{k+1})\| &= \|\mathrm{re}(w_1 w_2 \ldots w_{k-1})(w_k w_{k+1})\| \\ &\leqslant \prod_{j=1}^{k-1} (\|u_j\|+\|v_j\|)(\|u_n u_{n+1}-v_n v_{n+1}\|+\|u_{n+1} v_n+u_n v_{n+1}\|)\end{aligned}$$

which, combined with the inequality

$$\begin{aligned}&\|u_n u_{n+1}-v_n v_{n+1}\|+\|u_{n+1} v_n+u_n v_{n+1}\| \\ &\qquad\leqslant \|u_n\|\,\|u_{n+1}\|+\|v_n\|\,\|v_{n+1}\|+\|u_{n+1}\|\,\|v_n\|+\|u_n\|\,\|v_{n+1}\| \\ &\qquad= (\|u_n\|+\|v_n\|)(\|u_{n+1}\|+\|v_{n+1}\|)\end{aligned}$$

gives (5.5.2) for $n = k+1$. Inequality (5.5.2) being proved, we now pass to the proof of the fact that the function f is harmonic for each $\zeta \in C$. Fix $\zeta_0 = \alpha_0 + i\beta_0 \in C$. Employing the formal power series expansion

$$\frac{1}{\zeta - t} = \sum_{k=0}^{\infty} \frac{(\zeta-\zeta_0)^k}{(\zeta_0 - t)(t-\zeta_0)^k}$$

we obtain the relation

$$W_\xi = \mathrm{re}[(\zeta-t)^{-1}] = \sum_{k=0}^{\infty} \mathrm{re}\frac{(\zeta-\zeta_0)^k}{(\zeta_0 - t)(t-\zeta_0)^k}. \tag{5.5.3}$$

We shall show that equality (5.5.3) actually holds in the space W. In view of (5.5.2) we get the estimation:

$$\left\| W_\xi - \sum_{k=0}^{n} \mathrm{re}\frac{(\zeta-\zeta_0)^k}{(\zeta_0 - t)(t-\zeta_0)^k} \right\| = \left\| \mathrm{re}\left(\frac{1}{\zeta - t} - \sum_{k=0}^{n} \frac{(\zeta-\zeta_0)^k}{(\zeta_0 - t)(t-\zeta_0)^k} \right) \right\|$$
$$= \left\| \mathrm{re}\left[\left(\frac{\zeta-\zeta_0}{t-\zeta_0} \right)^{n+1} (\zeta - t)^{-1} \right] \right\| \leqslant C(\zeta)[B(\zeta)]^{n+1},$$

where

$$C(\zeta) = C(\alpha+\beta i) = \left\| \frac{\alpha - t}{(\alpha-t)^2+\beta^2} \right\| + \left\| \frac{\beta}{(\alpha-t)^2+\beta^2} \right\|,$$

$$B(\zeta) = \left\| \frac{(\alpha-\alpha_0)(t-\alpha_0) - \beta_0(\beta-\beta_0)}{(\alpha_0-t)^2+\beta_0^2} \right\| + \left\| \frac{(\alpha-\alpha_0)\beta_0 + (\beta-\beta_0)(t-\alpha_0)}{(\alpha_0-t)^2+\beta_0^2} \right\|.$$

From the inequality

$$B(\zeta) \leqslant |\alpha-\alpha_0| \left\| \frac{t-\alpha_0}{(\alpha_0-t)^2+\beta_0^2} \right\| + |\beta-\beta_0| \left\| \frac{\beta_0}{(\alpha_0-t)^2+\beta_0^2} \right\| +$$
$$+ |\alpha-\alpha_0| \left\| \frac{\beta_0}{(\alpha_0-t)^2+\beta_0^2} \right\| + |\beta-\beta_0| \left\| \frac{t-\alpha_0}{(\alpha_0-t)^2+\beta_0^2} \right\|$$

it follows that there exists a $\delta > 0$ such that $B(\zeta) < \frac{1}{2}$ for ζ in the rectangular neighbourhood of ζ_0 given by

$$|\alpha-\alpha_0| < \delta, \quad |\beta-\beta_0| < \delta; \tag{5.5.4}$$

hence

$$\left\| W_\zeta - \sum_{k=0}^{n} \mathrm{re}\left[\frac{(\zeta-\zeta_0)^k}{(\zeta_0 - t)(t-\zeta_0)^k} \right] \right\| \leqslant \frac{C(\zeta)}{2^n}.$$

This implies the convergence of series (5.5.3) for all ζ in the neighbourhood (5.5.4) of ζ_0. We thus have

$$f(\zeta) = \varphi(W_\zeta) = \sum_{k=0}^{\infty} \varphi\left(\mathrm{re}\left[\frac{(\zeta-\zeta_0)^k}{(\zeta_0 - t)(t-\zeta_0)^k} \right] \right),$$

whereby the right-hand series converges uniformly in the neighbourhood (5.5.4) as can be seen from the estimation

$$\left|\varphi\left(\operatorname{re}\frac{(\zeta-\zeta_0)^k}{(\zeta_0-t)(t-\zeta_0)^k}\right)\right| \leqslant \|\varphi\| \left\|\operatorname{re}\left[\frac{1}{\zeta_0-t}\left(\frac{\zeta-\zeta_0}{t-\zeta_0}\right)^k\right]\right\|$$

$$\leqslant \|\varphi\|\, C(\zeta_0)\, [B(\zeta)]^k \leqslant \|\varphi\| \frac{C(\zeta_0)}{2^k}.$$

Now write

$$\varrho_k(\zeta) = \operatorname{re}[(\zeta-\zeta_0)^k], \quad \sigma_k(\zeta) = \operatorname{im}[(\zeta-\zeta_0)^k],$$

$$\mu_k(t) = \operatorname{re}[(\zeta_0-t)^{-1}(t-\zeta_0)^{-k}], \quad \nu_k(t) = \operatorname{im}[(\zeta_0-t)^{-1}(t-\zeta_0)^{-k}].$$

Clearly $\mu_k, \nu_k \in W$ and ϱ_k, σ_k are harmonic functions of ζ, so

$$\varphi\left(\operatorname{re}\left[\frac{(\zeta-\zeta_0)^k}{(\zeta_0-t)(t-\zeta_0)^k}\right]\right) = \varphi(\varrho_k(\zeta)\mu_k - \sigma_k(\zeta)\nu_k)$$

is a harmonic function and consequently the function f is harmonic at each point of the complex plane C. It remains to verify that f is harmonic at infinity. The proof is similar to the former case. By the formal equality

$$\frac{1}{\zeta-t} = \sum_{k=0}^{n} \frac{t^k}{\zeta^{k+1}}$$

we obtain the relation

$$W_\zeta = \operatorname{re}[(\zeta-t)^{-1}] = \sum_{k=0}^{\infty} \operatorname{re}(\zeta^{-k-1}t^k). \tag{5.5.5}$$

The convergence of the right-hand series follows from the estimation:

$$\left\|W_\zeta - \sum_{k=0}^{n} \operatorname{re}(\zeta^{-k-1}t^k)\right\| = \|\operatorname{re}[(\zeta-t)^{-1}(\zeta^{-1}t)^{n+1}]\| \leqslant C(\zeta)\,[B(\zeta)]^{n+1},$$

where

$$C(\zeta) = C(\alpha+i\beta) = \left\|\frac{\alpha-t}{(\alpha-t)^2+\beta^2}\right\| + \left\|\frac{\beta}{(\alpha-t)^2+\beta^2}\right\|$$

and

$$B(\zeta) = \left\|\frac{\alpha t}{\alpha^2+\beta^2}\right\| + \left\|\frac{\beta t}{\alpha^2+\beta^2}\right\|.$$

We have

$$B(\zeta) \leqslant \|t\|\, [|\alpha|\,(\alpha^2+\beta^2)^{-1} + |\beta|\,(\alpha^2+\beta^2)^{-1}] \leqslant 2|\zeta|^{-1}\|t\|,$$

so $B(\zeta) < \frac{1}{2}$ for $|\zeta| > 4\|t\|$, whence

$$\left\|W_\zeta - \sum_{k=0}^{n} \operatorname{re}(\zeta^{-k-1}t^k)\right\| \leqslant 2^{-n-1}C(\zeta).$$

This implies the convergence of series (5.5.5) in a neighbourhood of infinity. Consequently the function $\zeta \to f(\zeta)$ is harmonic as a sum of a uniformly convergent series of harmonic functions; and this can be seen from the estimation

$$|\varphi(\mathrm{re}[\zeta^{-k-1}t^k])| \leqslant ||\varphi|| \, ||\mathrm{re}[t^{-1}(\zeta^{-k-1}t^{k+1})]||$$
$$\leqslant ||\varphi|| \, ||t^{-1}|| \, [B(\zeta)]^{k+1} \leqslant 2^{-k-1}||\varphi|| \, ||t^{-1}||$$

valid in the given neighbourhood in view of (5.5.2). The harmonicity of the function f on the closed plane $C \cup \{\infty\}$ leads to a contradiction, thus proving the assertion.

5.5.a. EXERCISE. Prove that if the norm of a real Banach algebra A satisfies the condition $||xy|| = ||x|| \, ||y||$ for all $x, y \in A$, then the algebra A is isomorphic to either of the three finite-dimensional linear fields.

5.6. REMARK. In the proof of Theorem 5.5 we did not make use of the completeness of the field in question; the only essential assumption was the fact that it has a submultiplicative norm. Even this is superfluous; it suffices to assume that the multiplication is separately continuous, since formula (2.4.1) defines then a submultiplicative norm in A. Of course, this norm need not be equivalent to the original one; but this is unessential for the proof.

§ 6. SOME REMARKS ON ALGEBRAS WITHOUT UNIT

In this section we deal with some concepts connected with algebras without unit: approximate identities, quasi-invertible elements and modular ideals.

6.1. DEFINITION. Let A be a Banach algebra. An *approximate identity* in A is a net of elements (δ_α) such that the relations

$$\lim_\alpha \delta_\alpha x = \lim_\alpha x\delta_\alpha = x$$

hold for every $x \in A$ and the net of norms $||\delta_\alpha||$ is bounded. In a similar way *left* and *right approximate identity* are defined.

6.2. EXAMPLE. Let $A = L_1(-\infty, \infty)$. We put

$$\delta_n(t) = \begin{cases} 2n & \text{for} \quad -\dfrac{1}{n} \leqslant t \leqslant \dfrac{1}{n}, \\ 0 & \text{for} \quad |t| > \dfrac{1}{n}. \end{cases}$$

We have $||\delta_n|| = 1$ and

$$x * \delta_n(t) = 2n \int_{-1/n}^{1/n} x(t-\tau)\, d\tau$$

whence

$$||x - x * \delta_n|| = \int_{-\infty}^{\infty} \left| 2n \int_{-1/n}^{1/n} x(t-\tau)\,d\tau - x(t) \right| dt$$

$$\leqslant 2n \int_{-1/n}^{1/n} \Big[\int_{-\infty}^{\infty} |x(t-\tau)-x(t)|\,dt \Big]\,d\tau.$$

As is well known (cf. e.g. Hewitt and Stromberg [1]), for every $x \in L_1(-\infty, \infty)$ we have

$$\lim_{\tau \to 0} \int_{-\infty}^{\infty} |x(t-\tau)-x(t)|\,dt = 0,$$

thus for any $\varepsilon > 0$ an integer N exists such that for $|\tau| < 1/N$ we have

$$\int_{-\infty}^{\infty} |x(t-\tau)-x(t)|\,dt < \varepsilon.$$

Hence it follows that

$$||x - x * y_n \delta_n|| \leqslant 2n \int_{-1/n}^{1/n} \varepsilon d\tau = \varepsilon$$

holds for $n > N$ and thus the sequence (δ_n) is an approximate identity in $L_1(-\infty, \infty)$.

6.2.a. Exercise. Find an approximate identity in $L_1(K)$.

6.2.b. Exercise. Find an approximate identity in the subalgebra $A = \{x \in C[0, 1]: x(0) = 0\}$.

6.3. Remark. If (δ_α) is an approximate identity in an algebra A with unit e, then $\lim \delta_\alpha = e$, by Definition 6.1 with e in place of x. Thus if an approximate identity is a divergent sequence (net), then A has no unit.

6.3.a. Exercise. Prove that the algebra $L_1(-\infty, \infty)$ has no unit.

6.3.b. Exercise. Prove that the algebras in Exercises 6.2.a and 6.2.b have no units.

The following theorem (a stronger version of the next corollary, called the *Cohen factorization theorem*) has many important applications.

6.4. Theorem. *Let $A \in \mathbf{B}$ and suppose that A has a left approximate identity (δ_α), i.e. $\lim_\alpha \delta_\alpha x = x$ for all $x \in A$ and $\sup_\alpha ||\delta_\alpha|| = k < \infty$. Let (x_n) be any sequence of elements in A tending to zero. Then there exists a sequence $(y_n) \subset A$ with $\lim y_n = 0$ and an element $z \in A$ such that*

(6.4.1) $$x_n = z y_n$$

for all n.

PROOF. Let $A_1 = A \oplus \{\lambda e\}$ and put

(6.4.2) $$t_s = (e+\varrho(e-\delta_1)) \dots (e+\varrho(e-\delta_s))$$

and

(6.4.3) $$z_s = (e+\varrho(e-\delta_s))^{-1} \dots (e+\varrho(e-\delta_1))^{-1},$$

where δ_i are suitably chosen elements of (δ_α) and ϱ is a positive scalar satisfying

(6.4.4) $$\frac{\varrho(1+k)}{1-\varrho(k+1)} < 1$$

and $0 < \varrho < (k+1)^{-1}$; the latter condition ensures that the z_i are well defined elements of A_1, since $||\varrho(e-\delta_\alpha)|| < 1$ (cf. proof of Theorem 5.1).

We now put

(6.4.5) $$y_{n,s} = t_s x_n$$

so that

(6.4.6) $$x_n = z_s y_{n,s}$$

for all n and s.

Since $\lim x_n = 0$, we can choose $N = N(s)$ in such a way that

$$||x_n|| \leqslant 2^{-s} ||t_s||^{-1}$$

for all $n \geqslant N(s)$, which implies

(6.4.7) $$||y_{n,s}|| \leqslant 2^{-s}$$

for all $n \geqslant N(s)$. Without loss of generality we may assume $N(s) \geqslant s$.

The sequence (δ_i) will be chosen in such a way that sequences (6.4.3) and (6.4.5) are convergent in A_1 with respect to s (with fixed n). To this end suppose that we have chosen elements $\delta_1, \dots, \delta_p \in (\delta_\alpha)$ is such a way that

(6.4.8) $$||y_{n,i} - y_{n,i+1}|| \leqslant 2(1+\varrho)^{-i}$$

for all $n < N(i)$ and $i = 1, 2, \dots, p-1$ and that

(6.4.9) $$||z_i - z_{i+1}|| \leqslant 2(1+\varrho)^{-i}$$

for $i \leqslant p-1$. We are looking for a suitable $\delta_{p+1} \in (\delta_\alpha)$ such that (6.4.8) and (6.4.9) hold for all $n < N(p)$ and $i = p$.

We have

(6.4.10) $$y_{n,p} - y_{n,p+1} = t_p x_n - t_p(e+\varrho(e-\delta_{p+1}))x_n = \varrho t_p(x_n - \delta_{p+1} x_n).$$

Since (δ_α) is a left approximate identity for A, and $x_n \in A$, we can choose δ_{p+1} to make the norm of the right-hand side of (6.4.10) as small as we want for the finite collection $x_1, x_2, \dots, x_{N(p)-1}$, so that (6.4.8) holds for all $n < N(p)$ and $i = p$, if, say, $\delta_{p+1} \in \{\delta_\alpha : \alpha > \alpha_0\}$.

In order to prove relation (6.4.9) consider the functional Φ defined on A_1 by the formula

$$\Phi(x+\lambda e) = \lambda$$

for $x \in A$, $\lambda \in C$. This is a continuous linear functional on A_1 having A as its zero set. We have $\Phi(e) = 1$ and for any $u, v \in A_1$, $u = x_u + \lambda_u e$, $v = x_v + \lambda_v e$, we have

$$\Phi(uv) = \Phi(\lambda_v x_u + \lambda_u x_v + x_u x_v + \lambda_u \lambda_v e) = \lambda_u \lambda_v = \Phi(u)\Phi(v)$$

since $\lambda_v x_u + \lambda_u x_v + x_u x_v \in A$. It follows also that $\Phi(u^{-1}) = (\Phi(u))^{-1}$ for any u invertible in A_1, since $1 = \Phi(e) = \Phi(uu^{-1}) = \Phi(u)\Phi(u^{-1})$. Such a functional Φ is called a *multiplicative linear functional* on A_1 and those functionals will be considered in Chapter II.

We can write now

$$z_{p+1} = (e + \varrho(e - \delta_{p+1}))^{-1} z_p$$

and so

(6.4.11)
$$\begin{aligned} z_p - z_{p+1} &= (e + \varrho(e - \delta_{p+1}))\, z_{p+1} - z_{p+1} \\ &= \varrho(e - \delta_{p+1})(e + \varrho(e - \delta_{p+1}))^{-1} z_p. \end{aligned}$$

Setting $R_s = z_s - \Phi(z_s) \cdot e$ we have $R_s \in A$ since $\Phi(R_s) = \Phi(z_s) - \Phi(z_s)\Phi(e) = 0$. We have also

(6.4.12)
$$\Phi(z_s) = [\Phi(e + \varrho(e - \delta_s))]^{-1} \ldots [\Phi(e + \varrho(e - \delta_1))]^{-1} = (1+\varrho)^{-s}$$

and so $R_s = z_s - (1+\varrho)^{-s} e$.

We can now rewrite (6.4.11) as

$$z_p - z_{p+1} = \varrho(e + \varrho(e - \delta_{p+1}))^{-1}(z_p - \delta_{p+1} z_p),$$

since $(e + \varrho(e - \delta_{p+1}))^{-1}$ and $e - \delta_{p+1}$ commute in A_1.

By the equality

$$z_p - \delta_{p+1} z_p = R_p - \delta_{p+1} R_p + (1+\varrho)^{-p}(e - \delta_{p+1})$$

we can write

(6.4.13)
$$\begin{aligned} z_p - z_{p+1} = {} & (R_p - \delta_{p+1} R_p)\varrho(e + \varrho(e - \delta_{p+1}))^{-1} + \\ & + (1+\varrho)^{-p} \varrho(e - \delta_{p+1})(e + \varrho(e - \delta_{p+1}))^{-1}. \end{aligned}$$

We can estimate the norms of both terms on the right-hand side of (6.4.13). We have

$$\begin{aligned} \|(e + \varrho(e - \delta_{p+1}))^{-1}\| &= \Big\| \sum_{n=0}^{\infty} (\varrho(e - \delta_{p+1}))^n \Big\| \\ &\leqslant \sum_{n=0}^{\infty} \varrho^n \|e - \delta_{p+1}\|^n \leqslant \frac{1}{1 - \varrho(k+1)} \end{aligned}$$

and so, by (6.4.4), the norm of the second term of the right-hand of (6.4.13) is less then

$$(1+\varrho)^{-p} \cdot \frac{\varrho(1+k)}{1 - \varrho(k+1)} < (1+\varrho)^{-p}.$$

The norm of the first term can be made as small as we want, e.g. smaller then $(1+\varrho)^{-p}$, if δ_{p+1} is suitably choosen in (δ_α), say, $\delta_{p+1} \in \{\delta_\alpha : \alpha > \alpha_1\}$, since (δ_α) is a left approximate identity in A and $R_p \in A$. By the choice of δ_{p+1} in $\{\delta_\alpha : \alpha > \max(\alpha_0, \alpha_1)\}$ we obtain both relations (6.4.8) and (6.4.9). Proceeding by induction we construct z_s and $y_{n,s}$ for all n and s in such a way that relations (6.4.8) and (6.4.9) are satisfied. From these relations it follows immediately that with fixed n there exist $y_n = \lim_s y_{n,s}$ and $z = \lim_s z_s$; $y_n \in A$ since $y_{n,s} \in A$, but $z_s \notin A$. However, by (6.4.12), $\Phi(z) = \lim \Phi(z_n) = \lim (1+\varrho)^{-n} = 0$ and so $z \in A$. Relation (6.4.1) is now a consequence of (6.4.6). It remains to show that $\lim y_n = 0$.

By (6.4.8) and the inequality $N(s) \geqslant s$ we have

$$||y_n - y_{n,N(s)}|| = \lim_{p\to\infty} \Big\| \sum_{i=N(s)}^{p} (y_{n,i+1} - y_{n,i}) \Big\| \leqslant \sum_{i=N(s)}^{\infty} 2(1+\varrho)^{-i}$$

$$= \frac{2}{1+(1+\varrho)^{-1}} (1+\varrho)^{-N(s)} \leqslant 2 \frac{1}{(1+\varrho)^{s-1}},$$

so by (6.4.7)

$$||y_n|| \leqslant 2^{-s} + \frac{2}{(1+\varrho)^{s-1}}$$

and $\lim ||y_n|| = 0$.

6.4.a. Exercise. Prove the following "right" version of Theorem 6.4: *If $A \in \mathbf{B}$ and A possesses a right approximate identity, then for any sequence $x_n \to 0$ of elements of A there is a sequence $y_n \to 0$, $y_n \in A$ and an element $z \in A$ such that $x_n = y_n z$ for $n = 1, 2, \ldots$*

6.5. Corollary (Cohen's factorization theorem). *Let $A \in \mathbf{B}$ and suppose that A possesses a left (or right) approximate identity. Then any element $x \in A$ can be written as a product $x = yz$ of two other elements $y, z \in A$, or, equivalently, $A^2 = A$.*

6.5.a. Exercise. Consider the space l_p, $1 < p < \infty$. Prove that it is a Banach algebra under coordinatewise algebra operations and show that it possesses an "unbounded approximate identity", i.e. a sequence (δ_n) of elements such that $\lim x\delta_n = \lim \delta_n x = x$ for all $x \in l_p$ but $||\delta_n|| \to \infty$. Show that $l_p^2 \subsetneqq l_p$ which proves that the boundedness assumption for an approximate identity is essential in the proof of Corollary 6.4′.

We now pass to the notion of quasi-regular elements otherwise called *quasi-invertible elements*. Let A be a Banach algebra without unit. Write $A_1 = A \oplus \{\lambda e\}$ and assume that an element $(e-x) \in A_1$ with $x \in A$ is invertible in A_1. It is easy to see that the inverse $(e-x)^{-1}$ must be of the form $e-y$ with $y \in A$. We thus have

$$(e-x)(e-y) = e.$$

This relation can be written in terms of the elements of A only. Namely, it is equivalent to the following one:

$$x+y-xy = 0.$$

6.6. Definition. We write

$$x \circ y = x + y - xy. \tag{6.6.1}$$

An element $x \in A$ is called *left-quasi-regular* or *left-quasi-invertible* provided there exists an element y called the *left-quasi-inverse of* x such that $x \circ y = 0$. *Right-quasi-regularity* is defined analogously. An element x is called *quasi-regular* (*-invertible*) provided it is both left- and right-quasi-regular. As pointed out below, it then follows that there exists an element $y \in A$ such that $x \circ y = y \circ x = 0$. This element, called the *quasi-inverse of* x, will be denoted by x°.

If $A_1 = A \oplus \{\lambda e\}$, then we can define a transformation T: $A \to A_1$ by $Tx = e - x$. The " $\circ$-multiplication" is transformed by T into usual multiplication, since $(e - x \circ y) = (e-x)(e-y)$. The transformation T carries the set of quasi-regular elements of A into the set of invertible elements of A_1 in a one-to-one way. It hence follows that the set of all quasi-regular elements is a group with respect to " $\circ$ ". The unit of this group is the zero element of A.

6.7. Lemma. *Let A be a Banach algebra; then all elements $x \in A$ with $\|x\| < 1$ are quasi-regular.*

The assertion follows immediately from the relation

$$x^{\circ} = -\sum_{n=1}^{\infty} x^n;$$

the verification is left to the reader.

The following theorem is an analogue of Theorem 5.1:

6.8. Theorem. *The set of all quasi-regular elements of a Banach algebra A is open.*

Proof. Write $A_1 = A \oplus \{\lambda e\}$. If x is a quasi-regular element of A, then there exists $x^{\circ} \in A$ with $x^{\circ} \circ x = 0$. For any $y \in A$ we have

$$\begin{aligned} x^{\circ} \circ y &= x^{\circ} + y - x^{\circ}y = x^{\circ} + y - x^{\circ}y - x^{\circ} \circ x = x^{\circ} + y - x^{\circ}y - x^{\circ} - x + x^{\circ}x \\ &= y - x - x^{\circ}y + x^{\circ}x = (e - x^{\circ})(y - x). \end{aligned}$$

Consequently

$$\|x^{\circ} \circ y\| \leqslant \|e - x^{\circ}\| \, \|y - x\|,$$

so, by Lemma 6.7, $x^{\circ} \circ y$ is quasi-regular for y close enough to x. Thus there exists a $z \in A$ such that $z^{\circ} \circ x^{\circ} \circ y = 0$, hence the element y is left-quasi-regular. Similarly we show that the elements sufficiently close to x are also right-quasi-reqular. So there exists a neighbourhood of x consisting of quasi-regular elements.

We now pass to the notion of a modular ideal, otherwise called regular.

6.9. Definition. An ideal $I \subset A$ is called a *left-* (*right-*) *modular ideal* provided there exists a *right-* (*left-*) *side unit modulo* I, i.e. an element e_I^r (e_I^l) such that $x - xe_I^r \in I$ ($x - e_I^l x \in I$) for any $x \in A$. Obviously, the elements e_I^r (e_I^l) occurring in this definition can be replaced be any elements e_I^r+m and e_I^l+m with any $m \in I$. Further note that if I is a proper ideal, then $e_I^r \notin I$ ($e_I^l \notin I$) and if $I \subset J$, then J also is a left- (right-) modular ideal with the unit e_I^r (e_I^l).

An ideal I is called a *modular ideal* if it is both right- and left-modular. By the above remarks it follows that for a modular ideal I there exists a *unit modulo* I, i.e. an element e_I such that $x - xe_I \in I$ and $x - e_I x \in I$ for any $x \in A$. For we have $e_I^l - e_I^l e_I^r \in I$ and $e_I^r - e_I^l e_I^r \in I$, so $e_I^l - e_I^r \in I$ and e.g. e_I^l can be taken as e_I.

6.9.a. Exercise. Prove that a two-sided ideal $I \subset A$ is modular if and only if the algebra A/I has a unit (the unit of A/I is the coset containing e_I).

6.10. Theorem. *Every proper modular ideal is contained in a maximal modular ideal.*

This is an immediate consequence of Kuratowski–Zorn's lemma.

6.11. Lemma. *Let A be a Banach algebra and let I be a left-modular ideal in A with the unit e_I^r. Then*

$$I \cap \{x \in A: \|e_I^r - x\| < 1\} = \emptyset.$$

Proof. If $\|e_I^r - x\| < 1$, then, by Lemma 6.6, $e_I^r - x$ is a quasi-regular element. Write $u = (e_I^r - x)^\circ$; thus $u + e_I^r - x - ue_I^r + ux = 0$ whence $e_I^r = (ue_I^r - u) + x - ux$. Suppose that $x \in I$; then we would have $e_I^r \in I$, contrary to the last remark in Definition 6.8.

This lemma implies the following two corollaries:

6.12. Corollary. *The closure of a proper modular ideal is a proper modular ideal.*

6.13. Corollary. *Every maximal modular ideal is closed.*

6.14. Remark. If an algebra A has a unit, then each of its ideals is modular (with the unit e). In particular, every maximal ideal in a Banach algebra with unit is closed.

§ 7. THE GROUP OF INVERTIBLE ELEMENTS OF A BANACH ALGEBRA

Let A be a Banach algebra with unit e. We know (Theorem 5.1) that the set G of all invertible elements of A is open and that it is a topological group with respect to the algebra multiplication (Theorem 5.3). We shall now examine the structure of this group. The set G, being an open subset of A, can be written as the union $G = \bigcup G_\alpha$ of its connected components G_α. Denote by G_0 that component which contains the unit e. It will be called the *principal component* of the group G of invertible elements.

7.1. THEOREM. *The principal component G_0 is a normal subgroup of G and is a maximal connected subgroup of G.*

PROOF. The second assertion is an immediate consequence of the first one, so we have to prove that G_0 is a normal subgroup of G. We first prove that G_0 is a subgroup of G. Let $a \in G_0$. The map $x \to ax$ is a homeomorphism of G onto itself, so it carries components onto components; in particular, aG_0 is a component of G. Similarly $G_0 a$ is a component of G. Since $a = ae \in aG_0$ and $a \in G_0 a$, then $aG_0 = G_0 = G_0 a$. Hence it follows that $a^{-1}G_0 = G_0 = G_0 a^{-1}$, so G_0 is a subgroup of G. Normality: for any $a \in G$ the map $x \to axa^{-1}$ is a homeomorphism of G onto itself which carries the component G_0 onto itself since $e \to aea^{-1} = e$. For any $a \in G$ we thus have $aG_0 a^{-1} = G_0$.

7.2. DEFINITION. Let A be a Banach algebra with unit e. Write

$$\exp x = \sum_{n=0}^{\infty} \frac{x^n}{n!}. \tag{7.2.1}$$

The series on the right-hand side converges absolutely in A: we have $\|\exp x\| \leqslant \exp\|x\|$. If $xy = yx$, then

$$\exp(x+y) = \exp x \exp y, \tag{7.2.2}$$

which can be verified as it is done with complex numbers. It hence follows that

$$\exp(-x)\exp x = e,$$

so the map $x \to \exp x$ maps A onto a subset $E \subset G$. A *logarithm of an element* $x \in A$ is any solution y of the equation

$$\exp y = x.$$

Thus the set E is precisely the set of those elements which have logarithms.

7.2.a. EXERCISE. Show by an example that (7.2.2) need not hold if $xy \neq yx$.

7.2.b. EXERCISE. Show that for $\|e-x\| < 1$ the series $y = -\sum_{n=1}^{\infty} \frac{1}{n}(e-x)^n$ converges and $\exp y = x$, so that $\{x \in A: \|x-e\| < 1\} \subset E$.

7.3. THEOREM. *The subgroup $G_1 \subset G$ generated by the set E is identical with the principal component G_0.*

PROOF. The set E is non-void (Exercise 7.2.b). Let $u \in E$, $u = \exp y$. Put

$$u(t) = \exp ty, \qquad t \in R. \tag{7.3.1}$$

We have $u(t+\tau) = u(t)u(\tau)$, $u(0) = e$ and $u(1) = u$; so $\{u(t)\}$ is a one-parameter subgroup of G.

Since

$$\|u(t)-u(\tau)\| = \|\exp \tau y(\exp[(t-\tau)y]-e)\| \leqslant \|\exp \tau y\| \, \|\exp[(t-\tau)y]-e\|$$

and

$$\|\exp[(t-\tau)y]-e\| = \left\| \sum_{n=1}^{\infty} \frac{(t-\tau)^n y^n}{n!} \right\|$$

$$\leqslant \sum_{n=1}^{\infty} \frac{|t-\tau|^n \|y\|^n}{n!} = \exp\|(t-\tau)\,y\|-1,$$

then

$$\|u(t)-u(\tau)\| \leqslant \|\exp \tau y\| [\exp(|t-\tau|\,\|y\|)-1].$$

Consequently, for any fixed τ we have $\lim_{t\to\tau} u(t)-u(\tau) = 0$ and the connectedness of the subgroup $\{u(t)\}$ follows. Thus $u(t) \in G_0$ for all $t \in R$, in virtue of Theorem 7.1. In particular, $u = u(1) \in G_0$, so $E \subset G_0$ and also $G_1 \subset G_0$. The proof will be complete if we show that G_1 is open and closed in G_0. Let $u_n \in G_1$, $n = 1, 2, \ldots$, and let $u_n \to u_0 \in G_0$. We shall show that $u_0 \in G_1$, i.e. G_1 is closed in G_0. Indeed, $u_n^{-1}u_0 \to e$, so $u_n^{-1}u_0 \in E \subset G_1$ for large n (Exercise 7.2.b), whence $u_0 = u_n(u_n^{-1}u_0) \in G_1$. Now we shall show that G_1 is open in G_0. Let $u_0 \in G_1$; if $\|u-u_0\| < \|u_0^{-1}\|^{-1}$, then $\|uu_0^{-1}-e\| < 1$ and $uu_0^{-1} \in E \subset G_1$ (Exercise 7.2.b), so $u = uu_0^{-1}u_0 \in G_1$. G_1 is thus open in G_0 which completes the proof.

7.4. Corollary. *If A is commutative, then $E = G_0$.*

7.5. Theorem (Nagumo). *An element $x \in A$ has a logarithm if and only if it is an element of a connected abelian subgroup of G.* In other words, $E = \bigcup \tilde{G}_\alpha$, *where $(\tilde{G}_\alpha)$ is the family of all connected abelian subgroups of G.*

Proof. If $u \in E$, then u is an element of the abelian subgroup $\{u(t)\} \subset G$ given by (7.3.1). Conversely, we shall show that any connected abelian subgroup $\tilde{G}_\alpha \subset G$ is contained in E. Let A_α denote a maximal commutative subalgebra of A containing the set $\tilde{G}_\alpha$. Let G_α^0 be the principal component of G_α, the group of invertible elements of A_α and let E_α be the analogue of the set E in the algebra A_α. We have $E_\alpha = G_\alpha^0$ by Corollary 7.4. But $\tilde{G}_\alpha \subset G_\alpha^0$, so all the elements of $\tilde{G}_\alpha$ have logarithms in A_α and automatically also in A. Hence $\tilde{G}_\alpha \subset E$.

7.6. Theorem (Lorch). *If A is a commutative complex Banach algebra with unit, then the group of its invertible elements is either connected or has infinitely many components.*

Proof. Let $x \in G$. We shall be done if we show that if $x \notin G_0$, then all the powers x^n (n an integer) are in different components of G. Suppose, on the contrary, that x^k and x^l with $k \neq l$ lie in the same component G_α. Then $G_\alpha x^{-l} = G_0$, so the element x^{k-l} belongs to the principal component G_0, i.e. $x^n \in G_0$ for some $n \neq 0$.

By Corollary 7.4 the element x^n has a logarithm, so there exists a $y \in A$ such that $x^n = \exp y$. Let $u = \exp(-y/n)$; $u \in G_0$ by Corollary 7.4. We have

$u^n = \exp(-y)$ and $(ux)^n = u^n x^n = \exp(y-y) = e$, which means that ux is an element of finite order. We shall prove that any such element belongs to G_0. For if $a^n = e$, $a \in G$, then, putting $a(\lambda) = \sum_{k=0}^{n} (\lambda-1)^k (\lambda a)^{n-k-1}$, $\lambda \in C$, we have

$$(\lambda^n - (\lambda-1)^n)e = [(\lambda a)^n - (\lambda-1)^n e] = (\lambda a - (\lambda-1)e)a(\lambda),$$

so $\lambda a - (\lambda-1)e \in G$ provided $\lambda^n \neq (\lambda-1)^n$. The function $\lambda \to \varphi_a(\lambda) = \lambda a - (\lambda-1)e$ is a continuous map of the complex plane C into A such that $1 \to a$, $0 \to e$. Further, we can find an arc $\{\lambda(t)\}$, $0 \leqslant t \leqslant 1$, in C connecting 0 with 1 and such that $[\lambda(t)]^n \neq [\lambda(t)-1]^n$. The map $t \to \varphi_a(\lambda(t))$ carries the segment $[0, 1]$ continuously into G and on account of Theorem 7.1 we have $\varphi_a(\lambda(t)) \in G_0$ for all $t \in [0, 1]$, so $a \in G_0$.

Thus $ux \in G_0$; and since $u \in G_0$, then $x = u^{-1}ux \in G_0$, contrary to the supposition that $x \notin G_0$. The assumption that two different powers of x lie in the same component led us to a contradiction.

7.7. PROBLEM. It is not known whether Lorch's theorem holds in non-commutative Banach algebras (the question is open for complex algebras since there exists a real Banach space X such that for the algebra $B(X)$ of all endomorphisms of X the group of invertible operators consists of two components, cf. Mitiagin and Edelstein [1]).

7.8. REMARK. The structure of the group G of invertible elements of a Banach algebra A does not determine uniquely the structure of the algebra. It can happen that the groups G_1 and G_2 of invertible elements of two non-isomorphic Banach algebras A_1 and A_2 are topologically isomorphic. Here is an example: Let $A_1 = C([-1, -\frac{1}{2}] \cup [\frac{1}{2}, 1])$ and $A_2 = C([0, 1] \cup \{2\})$. The algebras A_1 and A_2 are not isomorphic (this may be seen e.g. like this: the set $\{x \in A_2 \colon x(t) = 0$ for $t \in [0, 1]\}$ is a non-trivial minimal ideal in A_2, whereas there is no such ideal in A_1; see also Theorem 20.2). We shall show that the corresponding groups G_1 and G_2 are homeomorphically isomorphic. For a given element $x \in G_1$ define $y \in G_2$ as follows:

$$y(t) = \begin{cases} x(t-1) & \text{for} \quad t \in [0, \frac{1}{2}], \\ [x(-\frac{1}{2})/x(\frac{1}{2})]\, x(t) & \text{for} \quad t \in [\frac{1}{2}, 1], \\ x(\frac{1}{2})/x(-\frac{1}{2}) & \text{for} \quad t = 2. \end{cases}$$

It can be seen without much difficulty that the assignment $x \to y$ defines the desired isomorphism.

COMMENTS ON CHAPTER I

The symbol C.*n*. will denote comments concerning the *n*-th section. If comments to some section are omitted it means that we know of no results interesting enough to be included here. In the sequel all algebras are complex algebras, unless otherwise stated.

C.2. From Definition 2.1 it follows that if A is a topological algebra and (U_α) is a basis for its neighbourhoods of zero, then for every α there is a β such that

$$U_\beta^2 \subset U_\alpha. \tag{C.2.1}$$

On the other hand, if A is a topological linear space and an algebra, and if there is a basis of neighbourhoods of zero satisfying (C.2.1) (which means that the multiplication is jointly continuous at the origin in $A \times A$), then A is a topological algebra, i.e. multiplication is jointly continuous everywhere.

Similarly as for Banach spaces (cf. Appendix), the class of all Banach algebras is precisely the intersection of (complete) locally bounded algebras (shortly l.b. algebras) and locally convex algebras (shortly LC-algebras). In these comments we discuss mainly those two generalizations of Banach algebras, and, unless otherwise stated, we shall consider only complete algebras without further reference to this fact. In any l.b. algebra A there exists an equivalent p-homogeneous (cf. Appendix) pseudo-norm satisfying (2.5.1) and (2.5.2) if A possesses a unit element. This justifies another term for a (complete) l.b. algebra: a p-normed algebra will be which often used in the sequel. The possibility of introducing a submultiplicative pseudo-norm characterizes l.b. algebras with unit. We have namely

C.2.1. THEOREM. *Let A be an F-algebra with unit e; then the following conditions are equivalent*:

(i) *There is an equivalent metric, in A, $\varrho(x, y)$ satisfying*

$$\varrho(xy, 0) \leqslant \varrho(x, 0)\varrho(y, 0)$$

for all $x, y \in A$.

(ii) *A is a locally bounded algebra.*

(iii) *The topology of A can be given by means of a p-homogeneous pseudonorm $||x||$, $0 < p \leqslant 1$, satisfying conditions* (2.5.1) *and* (2.5.2).

The proof of the above theorem is similar to that of Theorem 2.4, and an analogue of Theorem 2.4 is also true if we replace a Banach algebra by a p-normed (= l.b.) algebra.

Concerning Remark 2.6 let us notice that there exist algebras, whose underlying vector spaces are normed spaces, in which the multiplication is separately continuous but not jointly continuous (e.g. $A = L_1(R) \cap L_2(R)$ with L_2-norm and convolution multiplication is such an example, which follows from a result in Żelazko [3]). It may be shown, however, that if A is an F-space and an algebra, and if multiplication in A is separately continuous, then it is jointly continuous.

Consider now the case of an LC-algebra A. According to formula (C.2.1) its topology can be given by a family $(||x||_\alpha)$ of seminorms such that for every α there is an index β such that

$$||xy||_\alpha \leqslant ||x||_\beta \, ||y||_\beta. \tag{C.2.2}$$

We have, however, no analogue of Theorem 2.4. The reason lies in the following

C.2.2. THEOREM. *Let X be an LC-space and let $B(X)$ denote the algebra of all continuous linear endomorphisms of X. Suppose that there is a topology on $B(X)$ making it a topological linear space such that the map $(T, x) \to Tx$ of $B(X) \times X$ into X is jointly continuous. Then X is a normed space.* (This result, due to the author, is unpublished.)

This makes some trouble in general LC-algebras. We do not know e.g. whether in an LC-algebra A with unit e one can introduce an equivalent system of seminorms giving its topology, such that (C.2.2) holds and $||e||_\alpha = 1$ for all α.

The situation is much simpler if we consider a special class of LC-algebras, namely so called locally multiplicatively convex, or shortly m-convex algebras, for which there exists an equivalent system of seminorms $||x||_\alpha$, such that

$$||xy||_\alpha \leqslant ||x||_\alpha \, ||y||_\alpha \tag{C.2.3}$$

for all $x, y \in A$. The class of all m-convex algebras will be also denoted by **M**. For any m-convex algebra A with unit e the seminorms $(||x||_\alpha)$, satisfying formula (C.2.3), can be chosen in such a way that $||e||_\alpha = 1$ for all α. This follows from a representation of a not necessarily complete m-convex algebra as a subalgebra of a Cartesian product of Banach algebras, or (for a complete algebra) as a projective limit of Banach algebras. Namely, if we set $N_\alpha = \{x \in A: ||x||_\alpha = 0\}$ we obtain a closed ideal in A and A/N_α, equipped with the norm $||x||_\alpha$ is a normed algebra. Denote its completion by A_α — it is a Banach algebra.

This procedure defines also a natural projection of A into A_α and so a map of A into the Cartesian product $\prod_\alpha A_\alpha$ which turns out to be an isomorphism into. Moreover, if we assume that together with a finite number $||x||_{\alpha_1}, \ldots, ||x||_{\alpha_k}$ of seminorms the family $(||x||_\alpha)$ contains also the maximum $||x||_\beta = \max\{||x||_{\alpha_1}, \ldots, ||x||_{\alpha_k}\}$, then the index set becomes a directed set if we write $\alpha \prec \beta$ when $||x||_\alpha$ is continuous with respect to $||x||_\beta$.

The theory of m-convex algebras is based on the following

C.2.3. THEOREM. *If A is a complete m-convex algebra, then it is the projective (inverse) limit of the above defined Banach algebras A_α.*

If A is a B_0-algebra (cf. Appendix), *then its topology can be given by means of a sequence of seminorms $||x||_i$ satisfying*

$$||x||_i \leqslant ||x||_{i+1}, \quad i = 1, 2, \ldots, \tag{C.2.4}$$

for all $x \in A$ and

$$||xy||_i \leqslant ||x||_{i+1} ||y||_{i+1}, \quad i = 1, 2, \ldots \tag{C.2.5}$$

If A is an m-convex B_0-algebra, then instead of (C.2.5) *we have*

$$||xy||_i \leqslant ||x||_i ||y||_i. \tag{C.2.6}$$

For the proofs and details the reader is referred to Michael [1] and Żelazko [8], unless indicated otherwise.

REMARK. Some authors (as e.g. Michael in [1]) call m-convex algebras of type B_0 *Fréchet algebras.*

C.3. In the sequel we make use of the following example of a p-normed algebra.

C.3.1. EXAMPLE. Denote by l_p the algebra of all two-sided complex sequences $x = (x_n)_{-\infty}^{\infty}$ such that

$$||x|| = \sum_{-\infty}^{\infty} |x_n|^p < \infty, \tag{C.3.1}$$

where p is a fixed real number satisfying $0 < p \leqslant 1$. This is a p-normed algebra if we introduce multiplication as the convolution

$$(x_n)(y_n) = \Big(\sum_{k=-\infty}^{\infty} x_{n-k}\, y_k\Big)_{n=-\infty}^{\infty}$$

since the p-homogeneous pseudonorm given by (C.3.1) is also submultiplicative. The algebra l_p possesses a unit element, namely the sequence $(\delta_{n,0})_{n=-\infty}^{\infty}$, where $\delta_{n,k}$ is the Kronecker symbol (cf. Example 3.5).

This example can be generalized by taking $l_p(G)$ on any discrete group G, with convolution multiplication. We cannot, however, have here non-discrete groups as can be seen by the following

C.3.2. THEOREM. *Let G be a locally compact group and let $0 < p < 1$. Then the space $L_p(G)$ (with respect to the left Haar measure) is an algebra under convolution if and only if G is a discrete group* (for proof cf. Żelazko [4]).

As examples of LC-algebras we can give two examples of B_0-algebras.

C.3.3. EXAMPLE. Let $\mathscr{E}$ be the algebra of all entire functions with pointwise algebra operations. Then it is an m-convex B_0-algebra with the topology of uniform convergence on compact sets (compact-open topology). This topology can be given by seminorms

$$||x||_n = \max_{|t| \leqslant n} |x(t)|.$$

C.3.4. EXAMPLE. Denote by L^ω the intersection $L^\omega = \bigcap_{p \geqslant 1} L_p[0, 1]$. It is a B_0-space with e.g. L_n-norms as seminorms, $n = 1, 2, \ldots$, and it is a (non-m-convex) B_0-algebra with coordinatewise algebra operations.

For details and more examples see Arens [1], Żelazko [7] and [14].

C.5. All results of this section are true for p-normed algebras. The Mazur–Gelfand theorem is true for m-convex algebras. It is also true for B_0-algebras (Arens [2], Żelazko [2] and [8]) but fails if we drop the assumption of completness (Williamson [1], Żelazko [8] and [14]), or metrizability (one can put a complete locally convex topology on the algebra of all germs of meromorphic functions at the origin of the complex plane).

The Mazur–Gelfand theorem is true for LC-algebras under the additional assumption of continuity of the inverse $x \to x^{-1}$.

It is worth remarking, that the Mazur–Gelfand type theorem fails also for F-algebras (an unpublished example is due to L. Waelbroeck).

Concerning the continuity of inversion we mention the following

C.5.1. THEOREM. *If $A \in \mathbf{M}_e$, then the operation of taking inverse $x \to x^{-1}$ is continuous on the group $G(A)$ of all invertible elements. An element $x \in A$ is invertible in A if and only if $\pi_\alpha(x)$ is invertible in A_α for every α* (cf. Theorem C.2.3; for proof cf. Michael [1]).

The set of all invertible elements of an m-convex algebra need not be open, it is not open e.g. in the algebra $\mathscr{E}$ (Example C.3.3).

The operation $x \to x^{-1}$ of taking the inverse need not be continuous in an LC-algebra, it is discontinuous e.g. in the algebra L^ω (Example C.3.4) (cf. Arens [1], Żelazko [8]).

For F-algebras the criterion for continuity of the inverse is given by the following

C.5.2. THEOREM. *Let A be an F-algebra with unit. Then the operation $x \to x^{-1}$ is continuous on the set $G(A)$ of all invertible elements in A if and only if $G(A)$ is a G_δ-set.* (cf. Banach [2], Żelazko [8].)

C.6. All results here are true also for p-normed algebras. The factorization theorem is true also for m-convex B_0-algebras. It is known in a stronger version than given here for Banach algebras which reads as follows:

C.6.1. THEOREM. *Let A be an m-convex B_0-algebra and suppose that there is a net $(\delta_\alpha) \subset A$ such that*

(i) $$||\delta_\alpha||_i \leqslant C$$

for all α and $i = 1, 2, \ldots$, where C is a constant not depending upon α and i;

(ii) $$\lim_\alpha ||\delta_\alpha x - x|| = 0$$

for all $x \in A$, $i = 1, 2, \ldots$

Then for any compact $K \subset A$ there exists an element $z \in A$ and a compact set $K' \subset A$ such that

$$K \subset zK'.$$

For the proof cf. Craw [1].

C.7. The given results are also true for p-normed algebras. Concerning the exponential function defined by (7.2.1) and other entire functions we start our comments with the following

C.7.1. DEFINITION. Let $\varphi(t) = \sum_0^\infty a_n t^n$ be an entire function of one complex variable t. We say that φ *operates in an algebra* $A \in \mathrm{LC}_e$ if for every $x \in A$ the series $\varphi(x) = \sum a_n x_n$ is convergent in A.

Obviously any entire function operates in any (complete) m-convex algebra. On the other hand, one can show (cf. e.g. Żelazko [8]) that if an entire function operates in the algebra L^ω (Example C.3.4), then it is a polynomial. The fact that not all entire functions operate in a non-m-convex B_0-algebra is true in general in the commutative case and we have the following

C.7.2. THEOREM. *Let A be a commutative B_0-algebra with unit e. Then A is an m-convex algebra if and only if all entire functions operate in A.*

It is an open question whether Theorem C.7.2 holds true in the non-commutative case.

C.7.3. THEOREM. *For every entire function φ there exists a commutative non-m-convex B_0-algebra A with unit in which the function φ operates.*

For the proofs of C.7.2 and C.7.3, cf. Mitiagin, Rolewicz and Żelazko [1].

Concerning the exponential function (7.2.1) we remark that there exists a non-m-convex B_0-algebra A and an element $x \in A$ such that $\exp x$ exists in A, but is non-invertible there. It may also occur that the only invertible elements in a B_0-algebra are scalar multiples of the unit (in an m-convex algebras there are always invertible elements of the form $\exp x$), a suitable example can be found in Mitiagin, Rolewicz and Żelazko [1].

We now quote a result related to C.7.2, but first we need the following definition.

C.7.4. DEFINITION. An algebra $A \in \mathrm{LC}_e$ is called a *Q-algebra* if the set $G(A)$ of all invertible elements in A is open.

C.7.5. THEOREM. *Let A be a commutative Q-algebra; then $A \in \mathbf{M}$ provided the map $x \to x^{-1}$ is continuous.*

This result was obtained first by Żelazko [8] for B_0-algebras (here the map $x \to x^{-1}$ is automatically continuous by Theorem C.5.2), and then by Turpin [3] for arbitrary (complete) LC-algebras.

C.7.6. THEOREM. *Let A be a B_0-algebra with unit e and suppose that φ and ψ are entire functions operating in A; then*

(i) *The superposition $\varphi \circ \psi$ also operates in A*

and

(ii) *The map $x \to \varphi(x)$ is a continuous map of A into itself.*

C.7.7. THEOREM. *Let A be a commutative B_0-algebra with unit; then there exists a transcendental entire function operating in A if and only if there exists an equivalent sequence of seminorms $(\|x\|_i)$ and a matrix $C_{i,n}$ of positive real numbers such that*

$$\|x_1 \dots x_n\|_i \leqslant C_{i,n} \|x_1\|_{i+1} \dots \|x_n\|_{i+1}$$

for all n-tuples of elements in A and all $i = 1, 2, \dots$

For the proofs of C.7.6 and C.7.7 cf. Żelazko [8].

CHAPTER II

Commutative Banach Algebras

All algebras in this chapter are commutative Banach algebras with unit, unless otherwise stated. Extending notation of Definition 2.2 we shall write $A \in \mathbf{B}$ if A is a Banach algebra (without any further assumptions) and $A \in \mathbf{BC}$ if A is commutative. Similarly, $A \in \mathbf{B}_e$ ($A \in \mathbf{BC}_e$) will mean that A is a (commutative) algebra with unit.

§ 8. MULTIPLICATIVE LINEAR FUNCTIONALS

8.1. DEFINITION. Let A be a Banach algebra with or without unit. A homogeneous and additive (not necessarily continuous) functional f defined on A is called a *multiplicative linear functional* if

(8.1.1) $$f(xy) = f(x)f(y) \quad \text{for all} \quad x, y \in A,$$

(8.1.2) $$f(x) \not\equiv 0.$$

The set of all multiplicative linear functionals will be denoted by $\mathfrak{M}'(A)$ or simply $\mathfrak{M}'$ when no confusion is likely.

8.1.a. EXERCISE. Prove that for a linear functional satisfying (8.1.1) condition (8.1.2) is equivalent to: $f(e) = 1$.

8.1.b. EXERCISE. Prove that if $f \in \mathfrak{M}'(A)$ and x is an invertible element in A, then $f(x^{-1}) = [f(x)]^{-1}$.

We are now going to establish the connection between the multiplicative linear functionals of an algebra A and its maximal ideals. It is easy to see that if $f \in \mathfrak{M}'$, then $M_f = \{x \in A : f(x) = 0\}$ is a maximal ideal in A because it is an ideal of codimension 1. On the other hand, if M is a maximal ideal in A, then it is closed by Remark 6.14; thus, in view of Theorem 5.4 and 1.13, the quotient algebra A/M is isomorphic to the complex field. The natural homomorphism $A \to A/M = C$ defines then a multiplicative linear functional in A. This functional will be denoted by f_M. As is easy to see, $f_{M_f} = f$ and $M_{f_M} = M$. So we have the following

8.2. THEOREM. *There exists a one-to-one correspondence between the maximal ideals of an algebra A and its multiplicative linear functionals; this correspondence is given by the relation*

$$f \leftrightarrow M_f \quad (M \leftrightarrow f_M).$$

On account of this correspondence we may call the elements of $\mathfrak{M}'(A)$ also *maximal ideals* thus identifying any maximal ideal M with the corresponding functional f_M. The set $\mathfrak{M}'$ will be called the *set of maximal ideals of A*.

Since the maximal ideals of an algebra A are closed and since they are precisely the zero sets of the functionals f, then, by the last assertion in 1.15, we have the following

8.3. Theorem. *Every multiplicative linear functional in an algebra A is continuous.*

8.3.a. Exercise. Show by an example that Theorem 8.3 is false in the case of non-complete normed algebras.

Now we prove the analogue of Theorem 8.2 for algebras without unit.

8.4. Theorem. *Let A be a commutative complex Banach algebra without unit. Then there exists a one-to-one correspondence between the multiplicative linear functionals of A and the maximal modular ideals of A, given by the relation*

$$f \leftrightarrow M_f \qquad (M \leftrightarrow f_M).$$

Proof. If M is a maximal modular ideal, then it is closed (Corollary 6.12) and the quotient algebra A/M is isomorphic to the field of complex numbers as a Banach algebra with unit and without proper idelas (1.13). The natural homomorphism $A \to A/M = C$ is a multiplicative linear functional as it was in 8.2. Conversely, if f is a non-zero multiplicative linear functional then the set $M = \{x \in A: f(x) = 0\}$ is a modular ideal. For then there exists an element e_M such that $f(e_M) = 1$ and for any $x \in A$ we have $f(x - e_M x) = f(x) - f(e_M)f(x) = 0$, i.e. $x - e_M x \in M$. The element e_M is thus a unit modulo M and the ideal M is a modular ideal. Evidently, it is a maximal ideal as a subspace of codimension 1.

8.4.a. Exercise. Apply the algebra of all entire functions of a complex variable with the norm $||\varphi|| = \max_{|z| \leqslant 1} |\varphi(z)|$ to prove that non-complete normed algebras can contain maximal ideals of infinite codimension. Hence it follows that Theorems 8.2 and 8.4 are not valid for non-complete algebras.

Hint. Consider the ideal consisting of such φ that $\varphi(n) = 0$ for $n > N$, N depending on φ. Show that if M is a maximal ideal containing this ideal, then the complex codimension of M is > 1; apply Frobenius' theorem formulated before Theorem 5.5.

Remark. For algebras without unit the symbol $\mathfrak{M}'$ will also denote the set of all multiplicative linear functionals or the set of all maximal modular ideals.

8.5. Corollary. *In commutative complex Banach algebras without unit multiplicative linear functionals are continuous.*

8.6. Corollary. *If A is a Banach algebra and* $f \in \mathfrak{M}'(A)$, *then* $||f|| = 1$.

Proof. First note that $|f(x)| \leqslant ||x||$ for any $x \in A$. For otherwise, if we had $|f(x_0)| > ||x_0||$ for some $x_0 \in A$, then, replacing x_0 by a suitable scalar multiple αx_0, we obtain $|f(\alpha x_0)| > 1 > ||\alpha x_0||$ whence $|f[(\alpha x_0)^n]| = |f(\alpha x_0)|^n \to \infty$ whereas

$\|(\alpha x_0)^n\| \leqslant \|\alpha x_0\|^n \to 0$, contrary to Corollary 8.5. Thus $\|f\| \leqslant 1$. On the other hand, $f(e) = 1$ and $\|e\| = 1$, so $\|f\| = 1$.

8.7. COROLLARY. *In every commutative Banach algebra with unit there exists at least one multiplicative linear functional.*

PROOF. If A is the complex field, then the identity is such a functional. If A is not a field, then there exists an element $x \in A$ which is not invertible and the set xA is an ideal which can be embedded in a maximal ideal. Thus the algebra A has a maximal ideal or, equivalently, has a multiplicative linear functional.

8.7.a. EXERCISE. Show by an example that Corollary 8.7 is not valid for algebras without unit, even under the assumption that the complexes are the field of scalars and that the algebras in question are commutative.

8.7.b. EXERCISE. Give an example of a non-commutative complex Banach algebra with unit and without multiplicative linear functionals.

8.7.c. EXERCISE. Give an example of a commutative real Banach algebra with unit and without real multiplicative linear functionals.

Multiplicative linear functionals are the basic tool in the problems concerning commutative Banach algebras. The above exercises show why it is more natural to investigate complex algebras. The reader will notice later that this is also due to the frequent use of methods of analytic functions in the theory of Banach algebras (as was already done in the proof of Theorem 5.4, see also the proof of Theorem 7.6).

8.8. DEFINITION. Let A be a Banach algebra, $x \in A$. We define a function $x\hat{}(M)$ on $\mathfrak{M}' = \mathfrak{M}'(A)$ by the formula

$$x\hat{}(M) = f_M(x). \tag{8.8.1}$$

This function will be called the *Gelfand transform* of the element $x \in A$. By Corollary 8.6,

$$\sup_{M \in \mathfrak{M}'} |x\hat{}(M)| \leqslant \|x\|. \tag{8.8.2}$$

Let $C_B(\mathfrak{M}')$ denote the Banach algebra of all bounded complex-valued functions defined on $\mathfrak{M}'$ with the norm $\|\varphi\| = \sup_{M \in \mathfrak{M}'} |\varphi(M)|$.

The following is an immediate consequence of Definition 8.8 and inequality (8.8.2):

8.9. THEOREM. *The map $h: x \to x\hat{}$ is a continuous homomorphism*(1) *of the Banach algebra A into the algebra $C_B(\mathfrak{M}')$. The image of the unit e of A (under this homomorphism) is the constant function $e\hat{}(M) \equiv 1$. Every multiplicative linear functional on A is of the form $f(x) = x\hat{}(M_0)$, where M_0 is a fixed point of $\mathfrak{M}'$.*

---

(1) This homomorphism will be called in the sequel the *Gelfand representation* or *transformation*.

8.10. REMARK. In the sequel $A^\wedge$ will denote the image of A under the homomorphism h: $A^\wedge = h(A) \subset C_B(\mathfrak{M}')$. $A^\wedge$ is a (possibly non-complete) normed algebra with the norm

$$\|x\|_s = \sup|x^\wedge(M)|. \qquad (8.10.1)$$

We conclude this section with the following theorem, important for further applications:

8.11. THEOREM. *An element $x \in A$ is invertible if and only if $x^\wedge(M) \neq 0$ for all $M \in \mathfrak{M}'$. In this case*

$$(x^{-1})^\wedge(M) = [x^\wedge(M)]^{-1}. \qquad (8.11.1)$$

PROOF. If $x^\wedge(M_0) = 0$, then $x \in M_0$ and x cannot be invertible. Conversely, if x is not invertible in A, then xA is a proper ideal in A, thus contained in some maximal ideal M_0. Clearly $x \in M_0$, so $x^\wedge(M_0) = 0$. Formula (8.11.1) follows immediately from Exercise 8.1.b.

REMARK. The symbol $\mathfrak{M}'$ is a preliminary one; it means that $\mathfrak{M}'$ is so far regarded merely as a set. This set will be later equipped with a topology (or even with two different topologies) and the resulting topological space will be denoted by $\mathfrak{M}$ or $\mathfrak{M}(A)$. Of course, all the theorems of this section remain true with $\mathfrak{M}$ instead of $\mathfrak{M}'$ since they do not involve the topological structure of $\mathfrak{M}$.

§ 9. THE GENERAL FORM OF MULTIPLICATIVE LINEAR FUNCTIONALS IN SPECIFIC BANACH ALGEBRAS

9.1. We shall prove that in the algebra $C(\Omega)$, Ω a compact space (Example 3.1), every multiplicative linear functional $f(x)$ has the form

$$f(x) = x(p) \qquad (9.1.1)$$

for some $p \in \Omega$.

We first show that there exists a $p_0 \in \Omega$ such that $x(p_0) = 0$ whenever $f(x) = 0$. Supposing that there is no such point so that for any $p \in \Omega$ an $x_p \in C(\Omega)$ exists with $f(x_p) = 0$ and $x_p(p) \neq 0$ and replacing, if necessary, $x_p(t)$ by $x_p(t)\overline{x_p(t)}$ we may assume $x_p(t) \geqslant 0$, $x_p(p) > 0$. The open sets $U_p = \{t \in \Omega : x_p(t) > 0\}$ cover the compact space Ω, so Ω is contained in a finite union $U_{p_1} \cup \ldots \cup U_{p_n}$; hence $y(t) = x_{p_1}(t) + \ldots + x_{p_n}(t) > 0$ and the element y is invertible in $C(\Omega)$. On the other hand, $f(y) = f(x_{p_1}) + \ldots + f(x_{p_n}) = 0$, which is impossible in virtue of Theorem 8.11. Thus the desired point p_0 exists. Let x be any element of $C(\Omega)$. If $f(x) = \alpha$, then $f(x - \alpha e) = 0$ whence $(x - \alpha e)(p_0) = 0$ and $x(p_0) = \alpha = f(x)$. Formula (9.1.1) is thus proved. As we see, the equality $A^\wedge = C(\Omega)$ holds and the Gelfand representation h is just the identity of the algebra $C(\Omega)$ onto itself.

9.2. We shall prove that in the algebra $L_1(-\infty, \infty)$ (Example 3.3) every multiplicative linear functional f is given by the formula

$$f(x) = \int_{-\infty}^{\infty} x(\tau) e^{ip\tau} d\tau \tag{9.2.1}$$

for some $p \in (-\infty, \infty)$.

Indeed, $f(x)$ is a continuous linear functional on the space L_1, so it is certainly of the form

$$f(x) = \int_{-\infty}^{\infty} k(t)x(t)\,dt,$$

where $k(t)$ is an essentially bounded measurable function defined on the real line. By the condition of multiplicativity we get the relation

$$\int_{-\infty}^{\infty} k(t)\Big(\int_{-\infty}^{\infty} x(t-\tau)y(\tau)d\tau\Big)\,dt = \int_{-\infty}^{\infty} k(t)x(t)\,dt \int_{-\infty}^{\infty} k(\tau)y(\tau)\,d\tau.$$

Interchanging the integrals on the left-hand side (the conditions of Fubini's theorem are clearly fulfilled) we obtain, after a suitable change of variables, the following equality:

$$\int_{-\infty}^{\infty}\int_{-\infty}^{\infty} x(t)y(\tau)k(t+\tau)\,dt\,d\tau = \int_{-\infty}^{\infty}\int_{-\infty}^{\infty} x(t)y(\tau)k(t)k(\tau)\,dt\,d\tau.$$

This equality holds for all functions $x, y \in L_1$, thus (see 1.21) $k(t+\tau) = k(t)k(\tau)$ almost everywhere. By the measurability of k it follows that k is of the form $k(t) = e^{\alpha t}$ (see 1.20). Since the function k is essentially bounded, α must be purely imaginary: $\alpha = ip$ with p real, which proves formula (9.2.1). Functionals (9.2.1) are obviously different for different p's, so there exists a one-to-one correspondence between the points of the real line and the elements of $\mathfrak{M}'$ and we may identify those two sets. From this point of view the functions $x\hat{}(p)$ are of the form

$$x\hat{}(p) = \int_{-\infty}^{\infty} x(t)e^{itp}\,dt.$$

These are continuous functions of the real variable p. In this case the Gelfand transform is precisely the Fourier transform of x.

9.3. We shall prove that in the algebra $L_1(0, 1)$ (Example 3.4) there are no multiplicative linear functionals.

Indeed, let $x \in L_1(0, 1)$ and let $x(t) = 0$ for $0 \leqslant t \leqslant \varepsilon$ for some ε, $0 < \varepsilon \leqslant 1$. Then $\underbrace{x * x * \ldots * x}_{n}(t) = 0$ for $0 \leqslant t \leqslant n\varepsilon$, whence $x^n = 0$ for sufficiently large n. Thus $[f(x)]^n = f(x^n) = f(0) = 0$ and $f(x) = 0$. But the set of those functions in $L_1(0, 1)$ which equal zero in some neighbourhood of $t = 0$ is dense and multiplicative linear functionals are continuous; consequently, there exist in $L_1(0, 1)$ no functionals satisfying (8.1.1) and (8.1.2).

9.4. We shall prove that in the algebra $l_1(-\infty, \infty)$ (Example 3.5) every multiplicative linear functional has the form

$$(9.4.1) \qquad f(x) = \sum_{n=-\infty}^{\infty} x_n e^{int}$$

for some real t, $0 \leqslant t < 2\pi$.

Consider the element $x_0 = (\delta_{1n})_{-\infty}^{\infty}$ (δ_{nk} is the Kronecker symbol). Let f be a multiplicative linear functional defined on $l_1(-\infty, \infty)$ and let $f(x_0) = \alpha$. The element x_0 is invertible (with the inverse $x_0^{-1} = (\delta_{n(-1)})$), so $\alpha \neq 0$ by Theorem 8.11. Since every element $x = (x_n)$ of $l_1(-\infty, \infty)$ can be written as the convergent series $x = \sum_{-\infty}^{\infty} x_n x_0^n$, then

$$(9.4.2) \qquad f(x) = \sum_{n=-\infty}^{\infty} x_n \alpha^n.$$

We shall show that $|\alpha| = 1$. Indeed, if we had $|\alpha| > 1$, then putting $y = (y_n)$, where $y_n = 0$ for $n > 0$ and $y_n = \alpha^n$ for $n \leqslant 0$, we obtain an element $y \in l_1(-\infty, \infty)$ with $f(y) = \sum_{-\infty}^{0} \alpha^n/\alpha^n$, which is impossible. If we had $0 < |\alpha| < 1$, then putting $z = (z_n)$, where $z_n = \alpha^{-n}$ for $n \geqslant 0$ and $z_n = 0$ for $n < 0$, we obtain an element $z \in l_1(-\infty, \infty)$ with $f(z) = \sum_{0}^{\infty} \alpha^n/\alpha^n$, which is also impossible. Thus $|\alpha| = 1$, so $\alpha = e^{it}$, $0 \leqslant t < 2\pi$ and formula (9.4.2) gives (9.4.1). Functionals (9.4.1) are evidently different for different $t \in [0, 2\pi)$. After an identification of the points 0 and 2π we may identify the set $\mathfrak{M}'$ with a circle. The transforms $x^\wedge(t)$ can be then regarded as absolutely convergent Fourier series

$$x^\wedge(t) = \sum_{n=-\infty}^{\infty} x_n e^{int};$$

these are continuous periodic functions defined on the real line or continuous functions defined on the unit circle of the complex plane.

9.4.a. Exercise. Prove that in the algebra $l_1^{\langle\omega_n\rangle}(-\infty, \infty)$ (Example 3.6) every multiplicative linear functional f has the form

$$(9.4.3) \qquad f(x) = \sum_{-\infty}^{\infty} x_n z^n,$$

where z is a complex number satisfying the condition $r \leqslant |z| \leqslant R$ with $r = \lim (\sqrt[n]{\omega_{-n}})^{-1}$, $R = \lim (\sqrt[n]{\omega_n})$.

9.4.b. Exercise. From Example 9.4 and Theorem 8.10 derive the following *Wiener theorem* (1):

---

(1) This proof is due to Gelfand and is one of the first applications of the theory of Banach algebras.

Let x be a periodic function of a real variable t, equal to its Fourier expansion

$$x(t) = \sum_{-\infty}^{\infty} x_n e^{int}$$

and let the sequence of coefficients x_n be absolutely summable. If, further, $x(t) \neq 0$ for each t, then the function $y(t) = 1/x(t)$ also has the Fourier expansion

$$y(t) = \sum_{-\infty}^{\infty} y_n e^{int}$$

with $\sum_{-\infty}^{\infty} |y_n| < \infty$.

9.5. We shall prove that in the algebra $\mathscr{A}$ (Example 3.9) every multiplicative linear functional has the form

$$f(x) = x(p) \tag{9.5.1}$$

with some p, $|p| \leqslant 1$.

In fact, let $p = f(x_0)$, where $x_0(\lambda) \equiv \lambda$. Certainly $|p| \leqslant 1$ since otherwise $x_0 - pe$ would be invertible in $\mathscr{A}$ and, by (8.11.1),

$$f([x_0-pe]^{-1}) = [f(x_0-pe)]^{-1} = [f(x_0)-p]^{-1},$$

which is impossible. For any polynomial $w(\lambda)$ we now have $f(w) = w(p)$, whence $f(x) = x(p)$ for any function $x \in \mathscr{A}$ since f is a continuous functional and the polynomials are dense in $\mathscr{A}$. Similarly as in Example 9.1 we have $\mathscr{A}^{\wedge} = \mathscr{A}$ and the Gelfand transformation is just the identity.

9.6. Let U be a bounded set in the space C^n. Let A be the algebra consisting of those continuous functions defined on U which can be approximated uniformly on U by polynomials of n variables. In other words, A is the closure in the norm $||x|| = \sup_{t \in U} |x(t)|$ of the algebra of all polynomials. It is a Banach algebra under this norm and pointwise multiplication of functions. We shall find the general form of a multiplicative linear functional in this algebra. For this purpose we introduce the notion of the polynomially convex hull of U (which will be employed also in the sequel, see Definition 17.9).

The *polynomially convex hull of a set U* is the set

$$P(U) = \bigcap_{w} \{(\lambda_1, \ldots, \lambda_n) \in C^n : |w(\lambda_1, \ldots, \lambda_n)| \leqslant \sup_{(\mu_1, \ldots, \mu_n) \in U} |w(\mu_1, \ldots, \mu_n)|\},$$

where the intersection is taken over all polynomials of n variables. It follows directly from this definition that the set $P(U)$ is compact and that every sequence of polynomials uniformly convergent on U is also uniformly convergent on the hull $P(U)$. It is also clear that $U \subset P(U)$. Hence it follows that the elements of $\mathscr{A}$ can be regarded as continuous functions defined on $P(U)$. Now, it can be seen without difficulty that every multiplicative linear functional f on A is of the form $f(x) = x(p)$ with some $p \in P(U)$. In fact, every multiplicative linear functional on the algebra of all polynomials of n variables is the evaluation at some point of

C^n. Continuous functionals (with respect to our norm) are precisely the points of $P(U)$, since for any other point we can find a sequence of polynomials convergent in A and divergent at this point. Now it remains only to make use of the fact that the polynomials are dense in A. Similarly as in the preceding example we have $A^\wedge = A$ and the Gelfand transformation is the identity map.

§ 10. THE MAXIMAL IDEAL SPACE

In this section we shall be concerned with the Gelfand representation $x \to x^\wedge$. The fact that all Gelfand transforms in the preceding section were continuous functions is not casual. We shall prove that the set $\mathfrak{M}'$ of maximal ideals of A can be given a topology such that the resulting topological space is compact (locally compact in the case of an algebra without unit) and the functions $x^\wedge \in A^\wedge$ are continuous. This topology, being due to Gelfand, is called the *Gelfand topology*. The set $\mathfrak{M}'$ with this topology will be denoted by $\mathfrak{M}$ (or $\mathfrak{M}(A)$ if the algebra in question is to be indicated).

10.1. DEFINITION. Let $M_0 \in \mathfrak{M}'(A)$. The *local base at the point* M_0 *of* $\mathfrak{M}$ is formed by the neighbourhoods

$$(10.1.1) \qquad V = V(M_0; x_1, \ldots, x_n; \varepsilon)$$
$$= \{M \in \mathfrak{M}: |x_i^\wedge(M) - x_i^\wedge(M_0)| < \varepsilon, \quad i = 1, 2, \ldots, n\},$$

where $x_1, \ldots, x_n \in A$, $\varepsilon > 0$. Any such neighbourhood is thus determined by a number $\varepsilon > 0$ and a finite number of elements $x_i \in A$.

10.2. THEOREM. *The space* $\mathfrak{M}$ *of all maximal ideals of an algebra A with the topology defined by the local bases* (10.1.1) *is a Hausdorff topological space.*

PROOF. We have to show the following facts: (i) the intersection of two neighbourhoods (10.1.1) of a point M_0 contains another such neighbourhood; (ii) if $M_1 \in V(M_0; x_1, \ldots, x_n; \varepsilon) \overset{\mathrm{df}}{=} V_0$, then there exists a neighbourhood V_1 of M_1 of the form (10.1.1) contained in V_0; (iii) for any two distinct points $M_1, M_2 \in \mathfrak{M}$ there exist disjoint neighbourhoods (10.1.1).

(i) We have

$$V(M_0; x_1, \ldots, x_n; \varepsilon) \cap V(M_0; y_1, \ldots, y_m; \delta)$$
$$\supset V(M_0; x_1, \ldots, x_n, y_1, \ldots, y_m; \min(\varepsilon, \delta)).$$

(ii) If $M_1 \in V(M_0; x_1, \ldots, x_n; \varepsilon) = V_0$, then

$$V(M_1; x_1, \ldots, x_n; \min_{i \leqslant n}[\varepsilon - |\hat{x_i}(M_1) - \hat{x_i}(M_0)|]) \subset V_0.$$

(iii) If $M_1 \neq M_2$, then there exists an $x_0 \in M_1 \setminus M_2$, so that $\hat{x_0}(M_1) \neq \hat{x_0}(M_2)$ and the neighbourhoods $V(M_1; x_0; \varepsilon)$, $V(M_2; x_0; \varepsilon)$ are disjoint for $\varepsilon = \frac{1}{4}|\hat{x_0}(M_1) - \hat{x_0}(M_2)|$.

10.3. Remark. The continuity of the functions $M \to x^{\wedge}(M)$ follows immediately from Definition 10.1. It is not difficult to see that the neighbourhoods (10.1.1) define the weakest topology making those functions continuous. If the elements of $\mathfrak{M}$ are regarded as linear functionals on A, then the topology of $\mathfrak{M}$ is precisely the relative topology defined on $\mathfrak{M}$ by the weak topology $\sigma(A', A)$.

10.4. Theorem. *The space $\mathfrak{M}(A)$ of a Banach algebra A is compact.*

Proof. With any element $x \in A$ we associate the closed disc on the complex plane $K_x = \{z \in C: |z| \leqslant ||x||\}$. Let $P = \prod_{x \in A} K_x$ be the product of the discs K_x with the product topology. By Tychonov's Theorem 1.9 P is a compact space. Let φ be a mapping $\mathfrak{M} \to P$ given by $M \to p$, where p_x, the x-coordinate of p, is equal $x^{\wedge}(M)$. Since, by (8.8.2), $p_x \in K_x$, so $p \in P$. Let $\mathfrak{M}_P$ denote the image of $\mathfrak{M}$ under φ. The mapping φ is one-to-one and, as can be easily verified, the relative topology of the subset $\mathfrak{M}_P \subset P$ is identical with the image topology induced by the Gelfand topology of $\mathfrak{M}$; thus φ is a homeomorphic injection. Now it suffices to prove that $\mathfrak{M}_P$ is a closed subset of P. Let $p \in P$ be a cluster point of $\mathfrak{M}_P$; we have to prove that $p \in \mathfrak{M}_P$. It should thus be shown that p is an image (under φ) of some multiplicative linear functional, i.e. that:

(i) $$p_{\alpha x+\beta y} = \alpha p_x + \beta p_y,$$

(ii) $$p_{xy} = p_x p_y,$$

(iii) $$p_e = 1.$$

We shall prove relation (ii) in detail; the other two are proved similarly. Consider the neighbourhood V of p in P defined by the x-, y-, xy-coordinates of p and a number $\varepsilon > 0$:

$$V(p; x, y, xy; \varepsilon) = \{q \in P: |q_x - p_x| < \varepsilon, |q_y - p_y| < \varepsilon, |q_{xy} - p_{xy}| < \varepsilon\}.$$

Since p is a cluster point of $\mathfrak{M}_P$, then V contains elements of $\mathfrak{M}_P$. Thus there exists an ideal $M \in \mathfrak{M}$ such that

$$|p_x - x^{\wedge}(M)| < \varepsilon, \quad |p_y - y^{\wedge}(M)| < \varepsilon, \quad |p_{xy} - x^{\wedge}(M) y^{\wedge}(M)| < \varepsilon.$$

It hence follows that

$$\begin{aligned} &|p_{xy} - p_x p_y| \\ &\quad = |p_{xy} - x^{\wedge}(M)y^{\wedge}(M) + x^{\wedge}(M)y^{\wedge}(M) - p_x y^{\wedge}(M) + p_x y^{\wedge}(M) - p_x p_y| \\ &\quad \leqslant |p_{xy} - x^{\wedge}(M)y^{\wedge}(M)| + |(x^{\wedge}(M) - p_x)y^{\wedge}(M)| + |p(_x y^{\wedge}(M) - p_y)| \\ &\quad \leqslant \varepsilon + 2\varepsilon ||x||. \end{aligned}$$

ε was arbitrary positive, so $p_{xy} = p_x p_y$. (i) and (iii) are proved similarly,

10.4.a. Exercise. Prove relations (i) and (iii).

10.5. Corollary. *The Gelfand transformation $x \to x^{\wedge}$ is a continuous homomorphism of A into the algebra $C(\mathfrak{M})$.*

We now pass to the case when A is an algebra without unit.

10.6. THEOREM. *Let A be a commutative complex Banach algebra without unit. Then the family of neighbourhoods* (10.1.1) *defines a topology τ in the set $\mathfrak{M}'$ of maximal modular ideals of A; the resulting space $\mathfrak{M} = (\mathfrak{M}', \tau)$ is locally compact. The image $A^\wedge$ of A under the Gelfand homomorphism $x \to x^\wedge$ is then contained in $C_0(\mathfrak{M})$, the algebra of all continuous functions on $\mathfrak{M}$ vanishing at infinity, i.e. such that for any $\varepsilon > 0$ the set $\{M \in \mathfrak{M}: |x^\wedge(M)| \geqslant \varepsilon\}$ is a compact subset of $\mathfrak{M}$.*

PROOF. If $\mathfrak{M}$ is empty, the assertion is obvious. Otherwise put $A_1 = A \oplus \{\lambda e\}$. Let $\mathfrak{M}_1 = \mathfrak{M}(A_1)$. By Theorem 10.4 $\mathfrak{M}_1$ is a compact space. Denote by F_∞ the element of $\mathfrak{M}_1$ given by $F_\infty(x+\lambda e) = \lambda$ for $x \in A$. If $f \in \mathfrak{M}$, then the functional F_f defined on A_1 by $F_f(x+\lambda e) = f(x)+\lambda$, $x \in A$ is in $\mathfrak{M}_1$ and $F_f \neq F_\infty$. Conversely, if $F \in \mathfrak{M}_1$, $F \neq F_\infty$, then the restriction f_F of F to A is an element of $\mathfrak{M}$. So we have a one-to-one correspondence $f \leftrightarrow F_f$ between the elements of the spaces $\mathfrak{M}$ and $\mathfrak{M}_1 \setminus \{F_\infty\}$. After the identification $f = F_f$ we have the set equality $\mathfrak{M}_1' = \mathfrak{M}' \cup \{F_\infty\}$.

From the obvious equality

$$V(M; x_1+\lambda_1 e, \ldots, x_k+\lambda_k e; \varepsilon) = V(M; x_1, \ldots, x_k; \varepsilon)$$

it follows that the topology of $\mathfrak{M}$ is identical with the relative topology of $\mathfrak{M}'$ regarded as a subset of $\mathfrak{M}_1$. Consequently the space $\mathfrak{M}$ is locally compact since it is obtained by removing a single point F_∞ from a compact space.

The last assertion follows from the fact that $x^\wedge(M_\infty) = 0$, where $x \in A \subset A_1$ and M_∞ is the ideal corresponding to the functional F_∞. This is because $|x^\wedge(M)| < \varepsilon$ in some neighbourhood of M_∞, i.e. off a compact subset of $\mathfrak{M}$.

10.7. COROLLARY. *The space $\mathfrak{M}(A_1)$, where $A_1 = A \oplus \{\lambda e\}$, is obtained from the space $\mathfrak{M}(A)$ by the one-point compactification (with the point M_∞). In the case when the space $\mathfrak{M}(A)$ is itself compact, i.e. when A has a unit* (cf. Exercise 10.7a), *the point M_∞ is an isolated point of $\mathfrak{M}(A_1)$.*

10.7.a. EXERCISE. Prove that if the space $\mathfrak{M}(A)$ is compact, then the algebra A has a unit.

10.8. DEFINITION. A subset $K \subset A$ is called a *system of generators* of a subalgebra A_0 of a Banach algebra A if A_0 is the smallest subalgebra of A containing the set K and the unit e of A; in symbols $A_0 = [K]$. If $[K] = A$, we say that A is *generated* by K. If A is generated by $K = \{x\}$, $x \in A$, then the element x is called a *generator* of A; the algebra A is then *singly generated.*

10.8.a. EXERCISE. Prove that $A = [K]$ if and only if the set of all finite sums $\sum a_{i_1, i_2, \ldots, i_k} x_1^{i_1} x_2^{i_2} \ldots x_k^{i_k}$ with $a_{i_1, i_2, \ldots, i_k} \in C$, $x_1, \ldots, x_k \in K$, is dense in A.

10.8.b. EXERCISE. Show that the algebra $C[0, 1]$ is singly generated.

10.8.c. EXERCISE. Prove that the algebra $C(K)$, K a circle, is not singly generated.

10.9. THEOREM. *If K is a system of generators for A, then the system of neighbourhoods* (10.1.1) *with $x_1, x_2, \ldots, x_n \in K$ is a base in the space $\mathfrak{M}(A)$.*

PROOF. We have to prove that every neighbourhood $U = U(M_0; x_1, \ldots, x_n; \varepsilon)$, $x_i \in A$, contains a neighbourhood $V = V(M_0; y_1, \ldots, y_k; \delta)$ with $y_i \in K$. By Exercise 10.8.a there exist polynomials p_i with complex coefficients and elements $y_{k,j} \in K$ such that

$$\|p_i(y_{i,1}, y_{i,2}, \ldots, y_{i,k_i}) - x_i\| < \tfrac{1}{3}\varepsilon, \quad i = 1, 2, \ldots, n.$$

By the continuity of the polynomials p_i and the Gelfand transforms of all occurring elements there exists a constant δ such that

$$\begin{aligned} V_0 &= V(M_0; y_{1,1}, y_{1,2}, \ldots, y_{1,k_1}, y_{2,1}, \ldots, y_{n,k_n}; \delta) \\ &\subset V(M_0; p_1(y_{1,1}, \ldots, y_{1,k_1}), p_2(y_{2,1}, \ldots, y_{2,k_2}), \ldots, p_n(y_{n,1}, \ldots, y_{n,k_n}); \tfrac{1}{3}\varepsilon). \end{aligned}$$

We shall show that $V_0 \subset U$ whence the assertion follows. Let $M \in V_0$. We have

$$|p_i[\hat{y}_{i,1}(M), \ldots, \hat{y}_{i,k_i}(M)] - p_i[\hat{y}_{i,1}(M_0), \ldots, \hat{y}_{i,k_i}(M_0)]| < \tfrac{1}{3}\varepsilon$$

so

$$\begin{aligned} |\hat{x}_i(M) - \hat{x}_i(M_0)| &\leqslant |\hat{x}_i(M) - p_i[\hat{y}_{i,1}(M), \ldots, \hat{y}_{i,k_i}(M)]| + \\ &\quad + |p_i[\hat{y}_{i,1}(M), \ldots, \hat{y}_{i,k_i}(M)] - \\ &\quad - p_i[\hat{y}_{i,1}(M_0), \ldots, \hat{y}_{i,k_i}(M_0)]| + \\ &\quad + |p_i[\hat{y}_{i,1}(M_0), \ldots, \hat{y}_{i,k_i}(M_0)] - \hat{x}_i(M_0)| \\ &< \tfrac{1}{3}\varepsilon + \tfrac{1}{3}\varepsilon + \tfrac{1}{3}\varepsilon = \varepsilon, \end{aligned}$$

i.e. $M \in U$.

10.10. COROLLARY. *If a Banach algebra A is countably generated, then the space $\mathfrak{M}(A)$ has a countable base and, by the compactness, is a complete metric space. This holds also in the case when A has no unit since then the space $\mathfrak{M}(A)$ is a G_δ subset of the complete metric space $\mathfrak{M}(A \oplus \{\lambda e\})$ and is therefore itself completely metrizable* (see Engelking [1]).

10.11. THEOREM. *If a Banach algebra A is finitely generated* (*by n generators*), *then the space $\mathfrak{M}(A)$ is homeomorphic to a compact subset of C^n.*

PROOF. Let $x_1, \ldots, x_n \in A$ be a system of generators for A. We shall prove that the mapping $M \to [\hat{x}_1(M), \ldots, \hat{x}_n(M)]$ is a homeomorphism of $\mathfrak{M}$ into C^n. In fact, this mapping is one-to-one, since the equalities $\hat{x}_i(M_1) = \hat{x}_i(M_2)$, $i = 1, 2, \ldots, n$, imply that $u\hat{\ }(M_1) = u\hat{\ }(M_2)$ for any $u = \sum a_{i_1, \ldots, i_n} x^{i_1} \ldots x^{i_n}$, $a_{i_1, \ldots, i_n} \in C$, i.e. $f_{M_1}(u) = f_{M_2}(u)$ for all u from a dense subset of A. Hence $f_{M_1}(x) = f_{M_2}(x)$ for all $x \in A$ and $M_1 = M_2$. Our mapping is obviously continuous and is therefore a homeomorphism as a continuous invertible mapping of a compact set $\mathfrak{M}$ onto a subset of C^n. Thus the maximal ideal space of an algebra $A = [x_1, \ldots, x_n]$ may be identified with the set

(10.11.1) $$\{[f(x_1), f(x_2), \ldots, f(x_n)] \in C^n : f \in \mathfrak{M}\}.$$

10.12. DEFINITION. Let A be a Banach algebra and let $x_1, \ldots, x_n \in A$. The set

$$\sigma(x_1, \ldots, x_n) = \{[\hat{x_1}(M), \ldots, \hat{x_n}(M)] \in C^n:\ M \in \mathfrak{M}\}$$

is called the *joint spectrum of the elements* $x_1, \ldots, x_n$. The set

$$\sigma(x) = \{x\hat{}(M) \in C:\ M \in \mathfrak{M}\}$$

is called the *spectrum of an element* $x \in A$. These are non-void compact sets. For an algebra without unit the spectrum $\sigma(x)$ denotes the spectrum of x in the algebra $A_1 = A \oplus \{\lambda e\}$. It follows from Theorem 10.11 that for a finitely generated algebra the joint spectrum of a system of generators is homeomorphic to the space $\mathfrak{M}$. The notion of the spectrum will be discussed in detail in Section 17; for now we just note that this notion coincides with that of the spectrum of the operator $y \to xy$, namely $\sigma(x) = \{\lambda \in C:\ (x-\lambda e)$ is not invertible in $A\}$; and this follows immediately from Theorem 8.11.

Now we prove a theorem which allows to determine the space $\mathfrak{M}$ for specific Banach algebras.

10.13. THEOREM. *Let A be a Banach algebra. Let $\mathfrak{M}_\tau$ denote the set $\mathfrak{M}'(A)$ with a topology τ satisfying the conditions*:

(i) *The functions $x\hat{}$ are continuous in $\mathfrak{M}_\tau$,*

(ii) *$\mathfrak{M}_\tau$ is a compact Hausdorff space.*

Then the spaces $\mathfrak{M}_\tau$ and $\mathfrak{M}$ are homeomorphic.

PROOF. The identity map of $\mathfrak{M}_\tau$ onto $\mathfrak{M}$ is continuous, since the inverse images of the neighbourhoods (10.1.1) are open in $\mathfrak{M}_\tau$ as can be seen from condition (i). Thus it is a homeomorphism as a continuous invertible mapping of a compact space.

The following exercises give applications of the above theorem.

10.13.a. EXERCISE. Show that $\mathfrak{M}(C(\Omega)) = \Omega$, Ω compact Hausdorff (Examples 3.1 and 9.1).

10.13.b. EXERCISE. Show that the maximal ideal space of the algebra $l_1(-\infty, \infty)$ is homeomorphic to the unit circle on the complex plane (Examples 3.5 and 9.4).

10.13.c. EXERCISE. Show that the space $\mathfrak{M}$ of the algebra $l_1^{\langle\omega_n\rangle}(-\infty, \infty)$ is homeomorphic to the ring $\{z \in C:\ r \leqslant |z| \leqslant R\}$, where $R = \lim \sqrt[n]{\omega_n}$, $r = \lim(\sqrt[n]{\omega_{-n}})^{-1}$ (see Exercise 9.4.a).

10.13.d. EXERCISE. Show that $\mathfrak{M}(\mathscr{A}) = \{z \in C:\ |z| \leqslant 1\}$ (Examples 3.9 and 9.5).

10.13.e. EXERCISE. Let U be a bounded subset of C^n and let $P(U)$ be the polynomially convex hull of U. Show that $\mathfrak{M}(A) = P(U)$, where A is the completion of the algebra of all polynomials of n variables with complex coefficients in the norm $\|x\| = \sup_{t \in U} |x(t)|$ (Example 9.6).

§ 11. THE STRUCTURE SPACE

In this section we consider another topology for the set $\mathfrak{M}'$ first introduced by Gelfand and Shilov and generalized by other authors. This topology is meaningful also in certain spaces of ideals in the case of non-commutative algebras. Here we shall restrict ourselves to the commutative case; readers interested in the general case are referred to the book by Rickart [1]. The topology is of a purely algebraic character. In this section we do not assume the existence of a unit in the algebras in question. This section may be without loss omitted by a reader interested in the applications of Banach algebras only.

First we give the following lemma:

11.1. LEMMA. *Every maximal modular ideal is a prime ideal* (1).

PROOF. Let M be a maximal modular ideal. We have to show that if $xy \in M$, then either x or y is an element of M. Assume that $x \notin M$. We shall show that $y \in M$. Write $M_y = \{z \in A: zy \in M\}$; it is an ideal containing M properly since $x \in M_y$. Hence $M_y = A$, consequently the unit e_M modulo M belongs to M_y. Thus $e_M y \in M$ and since $y - e_M y \in M$, then $y \in M$.

11.2. DEFINITION. We define a closure operator in the set $\mathfrak{M}'$ of maximal modular ideals of a commutative Banach algebra A (cf. Definition 6.9) setting

$$\overline{F} = \{M \in \mathfrak{M}': \bigcap F \subset M\} \tag{11.2.1}$$

for a set $F \subset \mathfrak{M}'$, where $\bigcap F$ stands for $\bigcap\limits_{M \in F} M$.

Such a definition requires a verification of the closure axioms:

(i) $$\overline{F_1 \cup F_2} = \overline{F}_1 \cup \overline{F}_2,$$

(ii) $$\overline{\{M\}} = \{M\}, \qquad M \in \mathfrak{M}',$$

(iii) $$\overline{\overline{F}} = \overline{F}.$$

We shall give now this verification. Let $F_1, F_2 \subset \mathfrak{M}'$ and put $I_1 = \bigcap F_1$, $I_2 = \bigcap F_2$ and $I_{1,2} = \bigcap (F_1 \cup F_2)$. Thus I_1, I_2 and $I_{1,2}$ are closed ideals in A. It is clear that

$$I_{1,2} = I_1 \cap I_2. \tag{11.2.2}$$

Let $M \in \overline{F}_i$, i.e. $I_i \subset M$, for $i = 1$ or $i = 2$. By (11.2.2) it is $I_{1,2} \subset M$ and so $M \in \overline{F_1 \cup F_2}$, what implies

$$\overline{F}_1 \cup \overline{F}_2 \subset \overline{F_1 \cup F_2}. \tag{11.2.3}$$

Let now $M \in \overline{F_1 \cup F_2}$, i.e. $I_{1,2} \subset M$, or $I_1 \cap I_2 \subset M$. We claim that either $I_1 \subset M$, or $I_2 \subset M$. If not, there is an element $x \in I_1 \setminus M$ and an element $y \in I_2 \setminus M$ and so $xy \in I_1 \cap I_2 \subset M$ while $x, y \notin M$ what is impossible by Lemma 11.1. Thus either $M \in \overline{F}_1$, or $M \in \overline{F}_2$ what implies $\overline{F_1 \cup F_2} \subset \overline{F}_1 \cup \overline{F}_2$. This relation together with (11.2.3) gives the verification of axiom (i).

The verification of (ii) is immediate.

(1) An ideal $I \subset A$ is called a *prime ideal* if the relations $xy \in I$, $x, y \in A$ yield either $x \in I$ or $y \in I$.

If $M \in \overline{F}$, then $\bigcap F \subset M$ and $M \cap \bigcap F = \bigcap F$. This implies $\bigcap \overline{F} \cap \bigcap F = \bigcap F$ and so $\bigcap \overline{F} \supset \bigcap F$. On the other hand, by (i) and (ii), it is $F \subset \overline{F}$. Consequently $\bigcap F \supset \bigcap \overline{F}$ what implies $\bigcap F = \bigcap \overline{F}$. Relation (iii) follows now from the fact that $M \in \overline{F}$ means $\bigcap F \subset M$ and $M \in \overline{\overline{F}}$ means $\bigcap \overline{F} \subset M$.

The closure (11.2.1) induces a topology in the set $\mathfrak{M}'$ called the *structure topology* of the maximal modular ideal space. The set $\mathfrak{M}'$ regarded as a topological space with the structure topology will be called the *structure space* of the algebra A and will be denoted by $\mathfrak{M}_s$ or $\mathfrak{M}_s(A)$.

11.3. REMARK. For a set $F \subset \mathfrak{M}'$ write

(11.3.1)
$$k(F) = \bigcap_{M \in F} M;$$

$k(F)$ is a subset of A called the *kernel of* F. For a set $X \subset A$ write

(11.3.2)
$$h(X) = \{M \in \mathfrak{M}' : X \subset M\},$$

called the *hull of* X.

In these symbols we have $\overline{F} = h(k(F))$. The structure topology is also called the *hull-kernel topology*.

The following theorems explain the algebraic character of the structure space.

11.4. THEOREM. *If I is a closed ideal of an algebra A, then the mapping $M \xrightarrow{\varphi} M/I$ is a homeomorphism of the closed subset $Q = \{M \in \mathfrak{M}_s(A) : I \subset M\}$ onto the space $\mathfrak{M}_s(A/I)$.*

PROOF. Note first that if e_M is a unit modulo $M \in \mathfrak{M}_s(A)$ and if $I \subset M$, then the coset $e_M + I$ is a unit modulo M/I in the algebra A/I and $M/I \in \mathfrak{M}_s(A/I)$. Thus φ maps Q into $\mathfrak{M}_s(A/I)$. It is not difficult to see that φ is one-to-one onto. In fact, for any $M' \in \mathfrak{M}_s(A/I)$ we have $\Phi^{-1}(M') \in Q$, where Φ is the natural homomorphism $A \to A/I$, and $\varphi(M) = \Phi(M)$ for all $M \in Q$. The set Q is closed in $\mathfrak{M}_s$ for if $M \in \overline{Q}$, then $\bigcap Q \subset M$ whence $I \subset M$ and $M \in Q$. In order to prove that φ is a homeomorphism we have to show that both φ and φ^{-1} carry closed sets into closed sets. So let $F = \overline{F} \subset Q$. We then have

$$\begin{aligned} \varphi(F) &= \{\varphi(M) : M \in F\} = \{\varphi(M) : \bigcap F \subset M\} \\ &= \{\varphi(M) : \bigcap \varphi(F) \subset \varphi(M)\} = \overline{\varphi(F)}. \end{aligned}$$

Conversely, if $\varphi(F) = \overline{\varphi(F)}$, then

$$\begin{aligned} F = \varphi^{-1}[\varphi(F)] = \varphi^{-1}[\overline{\varphi(F)}] &= \{\varphi^{-1}(M') : \bigcap \varphi(F) \subset M'\} \\ &= \{M : \bigcap F \subset M\} = \overline{F}. \end{aligned}$$

Now we shall prove a theorem in a sense dual to Theorem 11.4.

11.5. THEOREM. *If I is a closed ideal of A, then the mapping φ of the set $P = \{M \in \mathfrak{M}_s(A): I \not\subset M\}$ into $\mathfrak{M}_s(I)$ given by:*

$$\varphi: M \to M \cap I$$

is a homeomorphism of the open set $P \subset \mathfrak{M}_s(A)$ onto the space $\mathfrak{M}_s(I)$.

PROOF. The set P is open since its complement $Q = \{M \in \mathfrak{M}_s(A): I \subset M\}$ is closed by Theorem 11.3. The proof of the remaining assertions will be done in three steps. First we show that the mapping φ is one-to-one, then we show that it maps the set P onto the whole space $\mathfrak{M}_s(I)$ and at last we verify that it is a homeomorphism.

We start with the proof that φ is one-to-one. Let $M_1, M_2 \in P$, $M_1 \neq M_2$ and let $x, y \in A$ be elements such that $x \in M_1$, $x \notin M_2$, $y \in I$, $y \notin M_2$. We have $xy \in M_1 \cap I$ and, by Lemma 11.1, $xy \notin M_2$. So $xy \notin M_2 \cap I$ whence $\varphi(M_1) \neq \varphi(M_2)$ and φ is one-to-one.

It is easy to verify that for an $M \in P$ the ideal $M \cap I$ is a proper maximal ideal in I. It is the modularity of $M \cap I$ that requires a proof. Consider the multiplicative linear functional f_M defined on A. Since $I \not\subset M$, the restriction of f_M to I is an element of $\mathfrak{M}_s(I)$, or $M \cap I \subset \mathfrak{M}_s(I)$ (cf. Theorem 8.4). Thus $\varphi(P) \subset \mathfrak{M}_s(I)$.

Now we prove that $\varphi(P) = \mathfrak{M}_s(I)$. Let M' be any ideal in $\mathfrak{M}_s(I)$ and put $M = \{x \in A: xI \subset M'\}$. The set M is a proper ideal in A since $I^2 \not\subset M'$ on account of Lemma 11.1. We shall show that M is a maximal ideal in A. It suffices to show that M is a subspace of codimension 1, or, in other words, that, given an $x \notin M$, any element y of A may be written as a sum $y = m + \alpha x$ with $m \in M$ and $\alpha \in \boldsymbol{C}$. In view of the relation $x \notin M$ there exists an element $u \in I$ such that $ux \notin M'$. Let $y \in A$. If $y \in M$, then $y = 0x + m$, and if $y \notin M$, then there exists a $v \in I$ such that $vy \notin M'$. Clearly $u, v \notin M'$, so, by Lemma 11.1, $xvu, yvu \notin M'$. We thus have $f_{M'}(xuv) \neq 0 \neq f_{M'}(yvu)$, where $f_{M'}$ is the functional on I whose null-set is M'. Hence there exists a scalar $\alpha \neq 0$ such that $f_{M'}[(\alpha x - y)uv] = 0$ and $f_{M'}[(\alpha x - y)zuv] = 0$ for any $z \in I$. Since $uv \notin M'$, applying once more the preceding lemma, we see that $(\alpha x - y)z \in M'$ for any $z \in I$. Hence $\alpha x - y \in M$, so $y = \alpha x + m$, where $m \in M$; and M is a maximal ideal. The modularity of M: let $e_{M'} \in I$ be a unit modulo M' in the algebra I, so that $x - xe_{M'} \in M'$ for any $x \in I$; hence it follows that for any $x \in A$ we have $x - xe_{M'} \in M$ and $e_{M'}$ is a unit modulo M in the algebra A. Thus $M \in \mathfrak{M}_s(A)$. Since $I \cdot I \not\subset M'$, then $I \not\subset M$ and $M \in P$. Evidently $\varphi(M) = M'$ and consequently φ is a one-to-one mapping of the set P onto the whole space $\mathfrak{M}_s(I)$.

It remains to prove that φ and φ^{-1} are continuous in the structure topology, i.e. that the equality $F = \overline{F}$ is equivalent to the equality $\varphi(F) = \overline{\varphi(F)}$. But this is a consequence of the equivalence between the relations $M \supset \bigcap F$ and $\varphi(M) \supset \bigcap \varphi(F)$, where $F \subset P$; see the last argument in the proof of Theorem 11.4.

We now prove a theorem analogous to Theorem 10.4.

11.6. THEOREM. *If an algebra A has a unit, then the structure space $\mathfrak{M}_s(A)$ is compact.*

PROOF. We have to show that if $\{F_\lambda\}$ is a family of closed subsets of $\mathfrak{M}_s$ such that $\bigcap_\lambda F_\lambda = \emptyset$, then there exists a finite subfamily $F_{\lambda_1}, \dots, F_{\lambda_n}$ such that $\bigcap_{i=1}^n F_{\lambda_i} = \emptyset$. Put $I_\lambda = \bigcap_{M \in F_\lambda} M$ and denote by J the set of all finite sums $x_{\lambda_1} + x_{\lambda_2} + \dots + x_{\lambda_n}$ with $x_{\lambda_i} \in I_{\lambda_i}$. The set J is, of course, a (proper or not) ideal in A. If J were a proper ideal, then there would exist a maximal ideal $M \supset J$ whence $M \supset I_\lambda = \bigcap F_\lambda$ for all λ. Thus $M \in \overline{F}_\lambda = F_\lambda$ for all λ, contrary to the supposition that the intersection of all F_λ's is the void set. Consequently $J = A$, so there exist elements $x_{\lambda_1}, \dots, x_{\lambda_n}$ such that the unit $e = x_{\lambda_1} + \dots + x_{\lambda_n}$. We shall show that $\bigcap_{i=1}^n F_{\lambda_i} = \emptyset$. In fact, if M was a maximal ideal in $\bigcap_{i=1}^n F_{\lambda_i}$, then $I_{\lambda_i} \subset M$ would hold for $i = 1, 2, \dots, n$ so that $e \in M$. This contradiction concludes the proof.

11.7. REMARK. The functions $x^\wedge$ need not be continuous in the structure topology. Moreover, in view of Theorem 10.13 some of the functions $x^\wedge$ are certainly discontinuous provided the Gelfand topology and the structure topology do not coincide.

11.7.a. EXERCISE. Prove that the only continuous functions in the space $\mathfrak{M}_s(\mathscr{A})$ (Example 3.9) are constant functions.

The notions introduced in this section will be employed in Section 22 dealing with the so-called regular algebras.

§ 12. THE SPECTRAL NORM AND THE RADICAL

Let us turn again to the Gelfand representation $x \to x^\wedge$ of a commutative Banach algebra. As already seen (formula (8.8.2)),

$$\sup_{M \in \mathfrak{M}} |x^\wedge(M)| = \max_{M \in \mathfrak{M}} |x^\wedge(M)| \leqslant \|x\|.$$

The following concept plays an essential role in the theory of Banach algebras.

12.1. DEFINITION. Let $x \in A$. The *spectral norm of the element* x is the number

$$\|x\|_s = \max_{M \in \mathfrak{M}} |x^\wedge(M)| \leqslant \|x\|, \tag{12.1.1}$$

or, equivalently,

$$\|x\|_s = \max\{|\lambda| : \lambda \in \sigma(x)\}, \tag{12.1.2}$$

where $\sigma(x)$ is the spectrum of x (see Definition 10.12); this justifies the term, spectral norm.

12.2. THEOREM. *For every $x \in A$ there exists the limit*

$$\lim \sqrt[n]{\|x^n\|} = \|x\|_s. \tag{12.2.1}$$

PROOF. From inequality (8.8.2) it follows that for all $f \in \mathfrak{M}$ we have

$$|f(x)| = \sqrt[n]{|f(x^n)|} \leqslant \sqrt[n]{\|x^n\|},$$

whence

$$\|x\|_s \leqslant \liminf \sqrt[n]{\|x^n\|}.$$

It remains to show that $\limsup \sqrt[n]{\|x^n\|} \leqslant \|x\|_s$. For a fixed linear functiona $F \in A'$ write $\varphi_F(\lambda) = F[(e-\lambda x)^{-1}]$. It is not difficult to see that φ_F is a holo morphic function in the open disc $\{\lambda \in C: |\lambda| < 1/\|x\|_s\}$ (an entire function i the case $\|x\|_s = 0$). A direct calculation shows that $\varphi_F(\lambda) = \sum_{n=0}^{\infty} F(x^n)\lambda^n$, th series being convergent for all λ with $|\lambda| < 1/\|x\|_s$. For every such λ we thu have $\lim F(x^n)\lambda^n = \lim F[(\lambda x)^n] = 0$. This equality holds for all $F \in A'$, conse quently the sequence $(\lambda x)^n$ converges weakly to zero and is therefore bounded Thus there exists a constant C_λ such that $\|(\lambda x)^n\| \leqslant C_\lambda$ for $n = 1, 2, \ldots$, whenc $\|x^n\| \leqslant C_\lambda/|\lambda|^n$ and

$$\limsup \sqrt[n]{\|x^n\|} \leqslant \limsup \sqrt[n]{C_\lambda|\lambda|^{-n}} = |\lambda|^{-1}.$$

Since this equality holds for every λ with $|\lambda| < 1/\|x\|_s$, then

$$\limsup \sqrt[n]{\|x^n\|} \leqslant \|x\|_s.$$

12.3. COROLLARY. *Let $|||x|||$ be any norm equivalent to the original norm in A then there exists the limit*

$$\lim \sqrt[n]{|||x^n|||} = \|x\|_s$$

12.4. COROLLARY. *Let A_0 be a subalgebra of A and let $x \in A_0$. Let $\sigma(x)$ denot the spectrum of x in A and let $\sigma_0(x)$ denote the spectrum of x in A_0. Then th equality*

$$\max\{|\lambda|: \lambda \in \sigma(x)\} = \max\{|\lambda|: \lambda \in \sigma_0(x)\}$$

holds; or, equivalently,

$$\max_{M \in \mathfrak{M}(A)} |x^\wedge(M)| = \max_{M \in \mathfrak{M}(A_0)} |x^\wedge(M)|.$$

It follows from formula (12.1.1) that the spectral norm is a submultiplativ continuous pseudonorm in A and such that $\|e\|_s = 1$.

12.4.a. EXERCISE. Let A be a commutative Banach algebra. Put

$$K_1 = \{x \in A: \lim x^n = 0\},$$

$$K_2 = \{x \in A: \text{there exists an } n > 0 \text{ such that } \|x^n\| < 1\}.$$

Prove that $K_1 = K_2$ and that the set $K = K_1 = K_2$ is an open, convex, symmetric subset of A Show that the Minkowski functional $p(x)$ of the set K, i.e. $p(x) = [\sup\{|\lambda|: \lambda x \in K\}]^{-1}$ ($p(x) =$ if this supremum $= \infty$), is precisely the spectral norm of x.

12.5. COROLLARY. *Let A be any (not necessarily commutative) Banach algebra with a norm $\|\cdot\|$. Then for every $x \in A$ there exists the limit*

$$\lim \sqrt[n]{\|x^n\|} \stackrel{\text{df}}{=} \|x\|_s \leqslant \|x\|; \tag{12.5.1}$$

if, further, $xy = yx$, $x, y \in A$, then

$$\|x+y\|_s \leqslant \|x\|_s + \|y\|_s \tag{12.5.2}$$

and

$$\|xy\|_s \leqslant \|x\|_s \, \|y\|_s . \tag{12.5.3}$$

12.5.a. EXERCISE. Show by examples that if $xy \neq yx$, then relations (12.5.2) and (12.5.3) need not hold.

12.6. DEFINITION. Since in the non-commutative case x need not be a norm, the number $\lim \sqrt[n]{\|x^n\|}$ is referred to as the *spectral radius* of x.

12.6.a. EXERCISE. Prove that in any, possibly non-commutative Banach algebra A,

$$\|x^n\|_s = \|x\|_s^n . \tag{12.6.1}$$

The spectral norm is also characterized as follows: Let A be a Banach algebra with a norm $\|x\|$. An *admissible norm* will denote any norm $\|x\|_a$ equivalent to the norm $\|x\|$ and satisfying the conditions: $\|xy\|_a \leqslant \|x\|_a \|y\|_a$, $\|e\|_a = 1$. Let D denote the set of all admissible norms in A. Then the following theorem is true (Bohnenblust and Karlin [1]):

12.7. THEOREM. *Let A be a Banach algebra; then*

$$\|x\|_s = \inf\{\|x\|_a : \|\cdot\|_a \in D\}. \tag{12.7.1}$$

PROOF. According to (12.5.1) it suffices to show that if $x_0 \in A$ and $\|x_0\|_s < 1$, then there exists an admissible norm $\|\cdot\|_0$ such that $\|x_0\|_0 \leqslant 1$. Each element of A can be written (not uniquely) as a sum

$$x = \sum_{k=0}^{\infty} a_k x_0^k , \tag{12.7.2}$$

where $a_k \in A$, $x_0^0 = e$ and only finitely many a_k's are different from zero. We define

$$\|x\|_0 = \inf \sum_k \|a_k\|,$$

where the infimum is taken over all possible representations of x in the form (12.7.2). Clearly $\|x_0\|_0 \leqslant 1$; so the theorem will be proved if we show that $\|\cdot\|_0$ is an admissible norm.

If $y = \sum_k b_k x_0^k$, then $xy = \sum_k c_k x_0^k$, where $c_k = \sum_l a_{k-l} b_l$.

Hence

$$\|xy\|_0 \leqslant \sum_k \|c_k\| \leqslant \sum_{k,l} \|a_{k-l}\| \, \|b_l\| = \sum_k \|a_k\| \sum_l \|b_l\| .$$

Replacing the sums on the right-hand side by their infima taken with respect to all possible representations (12.7.2) of x and y, we get $||xy||_0 \leqslant ||x||_0 ||y||_0$. The subadditivity of the functional $||x||_0$ is proved in a similar way.

We have yet to prove that the norm $||x||_0$ is equivalent to the norm $||x||$ (and therewith we show that $||x||_0$ is in fact a norm, since so far it has been shown to be only a seminorm). Writing x as $x = x+0x_0+0x_0^2+\ldots$ we get $||x||_0 \leqslant ||x||$ for all $x \in A$. On the other hand, since $||x_0||_s < 1$, then by (12.2.1) the sequence $||x_0^n||$ is bounded, say, $||x_0^n|| \leqslant C$. It thus follows from (12.7.2) that $||x|| \leqslant C\sum_k ||a_k||$ whence $||x|| \leqslant C||x||_0$ and so the two norms are equivalent.

To see that $||e||_0 = 1$ it suffices to note that $||x||_s \leqslant ||x||_0 \leqslant ||x||$ for any $x \in A$. The second of these inequalities is proved in the preceding argument and the first one follows from (12.1.1).

We now pass to the notion of the radical of a Banach algebra. It follows from the definition of the spectral norm that the kernel $h^{-1}(0)$ of the Gelfand representation h is precisely the set

(12.8.1) $$\{x \in A: ||x||_s = 0\} = \bigcap_{M \in \mathfrak{M}} M.$$

12.8. Definition. The subset of an algebra A defined by formula (12.8.1) is called the *radical of* A and is denoted by $\operatorname{rad} A$. An element x belongs to the radical if and only if $\lim \sqrt[n]{||x^n||} = 0$. Such elements are also called *generalized nilpotent elements* by an analogy to *nilpotent elements*, i.e. those for which $x^n = 0$ holds with some integer n. Of course, every nilpotent element is in the radical. It is also evident that the radical is a closed ideal.

12.8.a. Exercise. Let $A_0 = \operatorname{rad} A$. Prove that either an integer N exists such that $x^N = 0$ for all $x \in A_0$ or there exist in A_0 elements x such that $x^n \neq 0$ for all n.

12.9. Theorem. *An element x of an algebra A is in the radical of A if and only if the inverse $(e+xy)^{-1}$ exists for all $y \in A$.*

Proof. If $x \in \operatorname{rad} A$, then for any $f \in \mathfrak{M}$ and any $y \in A$ we have

$$f(e+xy) = f(e)+f(x)f(y) = 1$$

by (12.8.1) and (12.1.1). Thus, in virtue of Theorem 8.11, the element $e+xy$ is invertible in A. If $x \notin \operatorname{rad} A$, then there exists a functional $f_0 \in \mathfrak{M}$ such that $\alpha = f_0(x) \neq 0$. Hence $f_0(e+(-e/\alpha)x) = 0$ and the element $e+(-e/\alpha)x$ is not invertible in A.

12.10. Example. Let $A = A_0 \oplus \{\lambda e\}$, where $A_0 = L_1(0, 1)$ (Example 3.4). We shall show that $\operatorname{rad} A = A_0$. In fact, put $x = x(\lambda) \equiv 1$. We have

$$x^n = \frac{\lambda^{n-1}}{(n-1)!},$$

so that x is a generator of A (Definition 10.8) because the polynomials are dense in A. We have further

$$\|x^n\| = \int_0^1 \frac{\lambda^{n-1}}{(n-1)!} \, d\lambda = \frac{1}{n!},$$

whence $\|x\|_s = \lim \sqrt[n]{1/n!} = 0$ and $x \in \operatorname{rad} A$. Since the radical is an ideal, then $xp(x) \in \operatorname{rad} A$, where p is any polynomial with complex coefficients. Every element $y \in A_0$ is a limit point of such polynomials, so the radical, being closed, contains A_0. On the other hand, no element $y = z + \lambda e$, with $z \in A_0$, $\lambda \neq 0$, belongs to the radical because the functional defined by $f(z+\lambda e) = \lambda$ is in $\mathfrak{M}(A)$ and $f(y) = \lambda \neq 0$. Hence $A_0 = \operatorname{rad} A$.

We now give (without proof) some facts concerning the radicals in non-commutative algebras. For the proofs the reader is referred to Rickart [1].

12.11. Let A be any Banach algebra; possibly non-commutative and without unit. Let X be any vector space over the field of the complexes. A *representation of A* in the space X is any homomorphism $x \to T_x$ of A into the algebra $L(X)$ of all linear operators $X \to X$. A subspace $X_0 \subset X$ is called an *invariant subspace of the representation* $x \to T_x$ provided $T_x X_0 \subset X_0$ for all $x \in A$. A representation is called *irreducible* if the only invariant subspaces are $\{0\}$ and X.

12.12. A two-sided ideal $I \subset A$ is called a *primitive ideal* if it is of the form $I = \{x \in A : xA \subset M\}$, where M is a maximal left-modular ideal in A. Such ideals are closed, hence also primitive ideals are closed. The following theorems hold (Rickart [1] (22.9) and (22.10)).

(i) *A two-sided ideal is primitive if and only if it is the kernel of an irreducible representation.*

(ii) *Every maximal two-sided modular ideal is primitive.*

(iii) *In commutative algebras primitive ideals are precisely the maximal modular ideals.*

12.13. The radical of any Banach algebra A, $\operatorname{rad} A$, is defined as the intersection of all primitive ideals. In the case when $A \neq \operatorname{rad} A$, the radical may also be defined as the intersection of all maximal left-modular ideals, or, equivalently, as the intersection of all such right ideals. The radical may also be defined by the formulas:

$$\begin{aligned}\operatorname{rad} A &= \{x \in A : \text{for every } \lambda \in \mathbf{C} \text{ the element } \lambda x + yx \text{ is quasi-regular for any } y \in A\} \\ &= \{x \in A : \text{for every } \lambda \in \mathbf{C} \text{ the element } \lambda x + xy \text{ is quasi-regular} \\ &\qquad\qquad \text{for any } y \in A\}.\end{aligned}$$

Let N denote the set of all generalized nilpotent elements of A. We always have $\operatorname{rad} A \subset N$ but it can happen that $\operatorname{rad} A \neq N$. An algebra A is called *semi-simple* if rad $A = \{0\}$. It can be proved that the algebra $B(X)$ of all bounded operators $X \to X$, X a Banach space, is always semi-simple. On the other hand, it is not difficult to construct in certain specific Banach spaces operators $T \neq 0$ such that $T^2 = 0$.

§ 13. SEMI-SIMPLE ALGEBRAS

13.1. Definition. A Banach algebra A is called *semi-simple* provided rad $A = \{0\}$. An algebra A is semi-simple if and only if the Gelfand homomorphism is an algebraic isomorphism.

We show in this section that the topology of semi-simple commutative Banach algebras is uniquely determined by the algebraic structure.

13.2. Theorem. *If Φ is an algebraic homomorphism of a Banach algebra A_0 into a semi-simple Banach algebra A_1, then Φ is continuous.*

Proof. It suffices to show that if $\Phi \not\equiv 0$, then the graph $\gamma = \{(x, \Phi(x)) \in A_0 \times A_1 : x \in A_0\}$ is closed in the product $A_0 \times A_1$ or, which is the same, that the conditions $x_n \to x_0$, $x_i \in A_0$, $\Phi(x_n) \to y_0$ imply the equality $y_0 = \Phi(x_0)$. Let $f \in \mathfrak{M}(A_1)$. Put $\tilde{f}(x) = f(\Phi(x))$. Since $\Phi \not\equiv 0$ and A_1 is semi-simple, then $\tilde{f} \in \mathfrak{M}(A_0)$. According to Theorem 8.3, $\tilde{f}(x_n) \to \tilde{f}(x_0) = f(\Phi(x_0))$. On the other hand, $f(\Phi(x_n)) \to f(y_0)$ whence $f(y_0) = f(\Phi(x_0))$, i.e. $f[\Phi(x_0) - y_0] = 0$ for every functional $f \in \mathfrak{M}(A_1)$. Since A_1 is semi-simple, then $y_0 = \Phi(x_0)$.

As a corollary we obtain

13.3. Theorem. *If Banach algebras A_1 and A_2 are algebraically isomorphic and one of them is semi-simple, then they are topologically isomorphic.*

Proof. If Φ is an isomorphism of A_1 onto A_2, then both A_1 and A_2 are semi-simple; thus, by the preceding theorem, both Φ and Φ^{-1} are continuous.

In particular, it follows that a given semi-simple commutative algebra can be normed in at most one way so as to be made a Banach algebra.

13.4. Corollary. *Every algebraic automorphism or endomorphism of a semi-simple Banach algebra is continuous.*

13.5. Corollary. *If the image $A\hat{}$ of a semi-simple algebra A is complete in the norm $\|\cdot\|_s$, then the norm in A is equivalent to the spectral norm.*

13.6. Exercises.

13.6.a. Show that Theorems and Corollaries 13.2-13.5 are no longer true without the assumption of semi-simplicity.

13.6.b. Show that the algebras in Examples 3.1, 3.2, 3.9 are semi-simple.

We now give (without proof) some facts concerning non-commutative algebras.

13.7. REMARK. Theorem 13.3 about the uniqueness of topology of a semi-simple Banach algebra is valid also in non-commutative algebras (Johnson [1]).

A particular case of this general statement is a much older result:

13.8. EIDELHEIT'S THEOREM. *The norm of the algebra $B(X)$ of all endmorphisms of a Banach space X is uniquely determined.*

Finally we give a theorem on representations of semi-simple, possibly non-commutative, Banach algebras. For the proof the reader is referred to the book by Rickart [1].

13.9. An algebra A is called a *primitive algebra* if $\{0\}$ is a primitive ideal (i.e. if A has a faithful irreducible representation). Let A_λ, $\lambda \in \Lambda$, be a family of Banach algebras. The set A of all functions $x(\lambda)$ defined on Λ and such that $x(\lambda) \in A_\lambda$ and $|||x||| = \sup_{\lambda \in \Lambda} ||x(\lambda)|| < \infty$ is also a Banach algebra; it is called the *direct sum* of the family $\{A_\lambda\}$. A *subdirect sum* of the family $\{A_\lambda\}$ is any closed subalgebra $A_0 \subset A$ such that the projection of A_0 onto the A_λ-axis is all of A_λ for each λ. The theorem states that every semi-simple Banach algebra is a subdirect sum of a family of primitive Banach algebras. The proof involves the space of all primitive ideals of a Banach algebra with a topology analogous to the structure topology.

§ 14. TOPOLOGICAL DIVISORS OF ZERO

14.1. DEFINITION. An element $x \neq 0$ of a Banach algebra A is called a *topological divisor of zero* if there exists a sequence $(x_n) \subset A$ such that $||x_n|| = 1$ and $\lim x_n x = 0$. In non-commutative algebras *left* and *right topological divisors of zero* are defined as well as *two-sided* ones, called just *topological divisors of zero*.

14.1.a. EXERCISE. Prove that a topological divisor of zero cannot be invertible.

14.1.b. EXERCISE. Prove that the condition $||x_n|| = 1$ in Definition 14.1 may be replaced by $||x_n|| \geqslant \delta > 0$.

14.2. THEOREM. *An algebra $A \in \mathbf{BC}_e$ has no topological divisors of zero if and only if it is isomorphic to the field of the complexes.*

PROOF. We shall prove that if A is not the complex field, then A has topological divisors of zero; the converse is obvious.

According to the supposition A contains an element x_0 which is not equal λe for any $\lambda \in C$. We shall show that if

$$\lambda_0 \in \overline{C \setminus \sigma(x_0)} \cap \sigma(x_0),$$

i.e. if λ_0 is in the boundary of the spectrum of x_0 (non-empty by 10.12), then the element $x_0 - \lambda_0 e$ is a topological divisor of zero in A. Let us take any sequence of numbers $\lambda_n \notin \sigma(x_0)$ which converges to λ_0. The elements $x_0-\lambda_n e$ are invertible in A in virtue of Theorem 8.11. Put

$$y_n = (x_0-\lambda_n e)^{-1}/\|(x_0-e)^{-1}\|.$$

We have $\|y_n\| = 1$ and

$$\|(x_0-\lambda_n e)^{-1}\| \geqslant \|(x_0-\lambda_n e)^{-1}\|_s = \max_{f\in\mathfrak{M}} |f(x_0)-\lambda_n|^{-1} \geqslant |\lambda_0-\lambda_n|^{-1} \to \infty .$$

Hence

$$\begin{aligned}\|(x_0-\lambda_0 e)y_n\| &\leqslant \|(x_0-\lambda_n e)y_n\|+|\lambda_0-\lambda_n|\,\|y_n\| \\ &= \|(x_0-\lambda_n e)^{-1}\|^{-1}+|\lambda_0-\lambda_n| \to 0.\end{aligned}$$

14.3. Corollary. *The above theorem is valid also for non-commutative Banach algebras with unit. In fact, any such algebra contains a commutative subalgebra non-isomorphic to the complex field; and this subalgebra has topological divisors of zero by the above theorem.*

We now prove a more general theorem (Żelazko [10]).

14.4. Theorem. *Let A be any complex Banach algebra (no commutativity or existence of a unit is assumed). Then either A has topological divisors of zero or A is isomorphic to the field of complex numbers.*

Proof. It can be easily verified that if every commutative subalgebra of A is isomorphic to the complex field, then so is A itself (as follows e.g. from the remark preceding Theorem 5.5). Without loss of generality we may thus assume that A is commutative. We have to show that if A is not isomorphic to the complex field, then it has topological divisors of zero. In view of Theorem 14.2 we may assume that A has no unit. Put $A_1 = A\oplus\{\lambda e\}$; A_1 has topological divisors of zero by the preceding theorem. Thus there exist elements $x_n \in A$ and numbers $\lambda_n \in \boldsymbol{C}$, $n = 0, 1, 2, \ldots$, such that $\|x_n\|+|\lambda_n| = 1$, $n = 0, 1, 2, \ldots$, and $\lim_n \|(x_0+\lambda_0 e)(x_n+\lambda_n e)\| = 0$. Clearly $x_0 \neq 0$. The sequence (λ_n), being bounded, may be assumed to be convergent: $\lambda_n \to \lambda$. Hence $\lim_n \|(x_0+\lambda_0 e)(x_n+\lambda e)\| = 0$. A has no unit, so $\lambda_0\lambda = 0$. The case $\lambda_0 = \lambda = 0$ means that the algebra A itself has topological divisors of zero. We consider the two remaining cases.

1° Let $\lambda_0 = 0$, $\lambda \neq 0$. Then $x_0(x_n+\lambda e) \to 0$. Suppose first that the sequence (x_n) is convergent, $x_n \to y$. Then $x_0(y+\lambda e) = 0$ and there exists a $z \in A$ such that $u = z(y+\lambda e) \neq 0$ since otherwise the element $-y/\lambda$ would be a unit of the algebra A, contrary to the assumption. Thus $u \in A$ and $ux_0 = 0$, so that A has divisors of zero and automatically also topological divisors of zero. If the sequence (x_n) is divergent, then there exists a $\delta > 0$ and a sequence of indices $k_n \to \infty$ such that $\|x_n-x_{k_n}\| \geqslant \delta$. Then $x_0(x_n-x_{k_n}) = x_0(x_n+\lambda e)-x_0(x_{k_n}+\lambda e) \to 0$ and x_0 is a topological divisor of zero (see Exercise 14.1.b).

2° Now assume that $\lambda_0 \neq 0$ and $\lambda = 0$, i.e. $(x_0+\lambda_0 e)x_n \to 0$. Similarly as previously there exists a $y \in A$ such that $u = (x_0 + \lambda_0 e)y \neq 0$. Thus $u \in A$, $ux_n \to 0$, $||x_n|| \to 1$ and A again has topological divisors of zero.

14.4.a. EXERCISE. Prove, with use of Theorem 5.5, the following extension of Theorem 14.4: Let A be any Banach algebra over the field of real numbers; then either A has topological divisors of zero or A is isomorphic to one of the three finite-dimensional fields over the reals.

Topological divisors of zero can be characterized as those elements which have no inverse in any superalgebra of the algebra in question.

14.5. DEFINITION. Let $A \in \mathbf{BC}_e$. An algebra $A_1 \in \mathbf{BC}_e$ is called an *extension* or a *superalgebra* of A if A_1 contains a subalgebra containing the unit e and isomorphic to A. An extension A_1 is called an *isometric extension* if the subalgebra in question is isometric to A. In each case we shall write $A \subset A_1$.

14.6. DEFINITION. An element $x \in A \in \mathbf{BC}_e$, $x \neq 0$, is called *permanently singular* if x is not invertible in any superalgebra $A_1 \supset A$.

14.7. THEOREM. *An element $z \in A \in \mathbf{BC}_e$ is permanently singular if and only if it is a topological divisor of zero.*

PROOF. A topological divisor of zero in A is also a topological divisor of zero in any extension $A_1 \supset A$, so, by Exercise 14.1.a, it is not invertible there—hence is permanently singular.

Now assume that z is not a topological divisor of zero and that $z \neq 0$. We have to construct a superalgebra $A_1 \supset A$ such that z is invertible in A_1. Let $\tilde{A}$ denote the algebra whose elements are formal power series

$$\tilde{x}(t) = \sum_{n=0}^{\infty} x_n t^n$$

with $x_n \in A$, such that

$$||\tilde{x}|| = \sum_{n=0}^{\infty} ||x_n|| < \infty .$$

It is not difficult to see that $\tilde{A}$ is a Banach algebra under the Cauchy multiplication of power series and that $\tilde{A}$ is a superalgebra of A; an isometric embedding is given by the assignment

$$x \leftrightarrow \tilde{x}(t) \equiv x$$

(i.e. an element $x \in A$ corresponds to a "constant" series $\tilde{x} \in \tilde{A}$).

By assumption we have $\delta = \inf_{||y||=1} ||zy|| > 0$ in A. Without loss of generality we may assume $\inf_{||y||=1} ||zy|| \geqslant 1$ (replacing z by z/δ), or, which is the same,

$$(14.7.1) \qquad ||zy|| \geqslant ||y|| \quad \text{for all } y \in A.$$

Consider the ideal $I = (e - z \cdot t)\tilde{A} \subset \tilde{A}$ and the quotient algebra $A_1 = \tilde{A}/I$. If

$[x(t)]$ denotes the coset of an element $x(t) \in \tilde{A}$, then clearly $[e] = [z]\,[t]$ whence $[z] \in G(A_1)$ provided that $[e] \neq 0$, i.e. that $I \neq \tilde{A}$. The theorem will thus be proved if we show that the inclusion $x \to [x]$ is an isometry of A into A_1. But this fact follows immediately from the following estimates:

$$\|x\| = \inf_{y(t)\in\tilde{A}} \|x+(e-zt)y(t)\| \leqslant \|x\|; \tag{14.7.2}$$

on the other hand we have, by (14.7.1),

$$\Big\|x-(e-zt)\sum_{n=0}^{\infty} y_n t^n\Big\| = \Big\|x-y_0+\sum_{n=1}^{\infty}(zy_{n-1}-y_n)t^n\Big\|$$

$$= \|x-y_0\|+\sum_{n=1}^{\infty}\|zy_{n-1}-y_n\| \geqslant \|x-y_0\|+\sum_{n=1}^{\infty}(\|zy_{n-1}\|-\|y_n\|)$$

$$\geqslant \|x-y_0\|+\sum_{n=1}^{\infty}(\|y_{n-1}\|-\|y_n\|) = \|x-y_0\|+\|y_0\| \geqslant \|x\|.$$

Hence

$$\|[x]\| \geqslant \|x\|. \tag{14.7.3}$$

Thus, by (14.7.2) and (14.7.3), $\|[x]\| = \|x\|$ holds for all $x \in A$, i.e. A_1 is an isometric extension of A such that the element z is invertible in A_1.

The above theorem is due to Arens [6].

14.7.a. Exercise. Prove that the algebra $\tilde{A}$ constructed in the proof of Theorem 14.7 is a Banach algebra and find its maximal ideal space.

As we have seen, each element which is not a topological divisor of zero is invertible in some superalgebra. Generally there exists no common extension for all such elements. An algebra A can, however, be embedded in a superalgebra in which every topological divisor of zero in A is actually a divisor of zero.

14.8. Theorem. *Let A be any Banach algebra (no commutativity or existence of a unit is assumed). Then there exists such a Banach algebra $\tilde{A}$ that A is isomorphic to a subalgebra of $\tilde{A}$ and every topological divisor of zero in A is a divisor of zero in A.*

Proof. Let A' be the algebra consisting of all sequences $x = (x_n) \subset A$ such that $|||x||| = \limsup \|x_n\| < \infty$. Let N denote the ideal $N = \{x \in A' \colon |||x||| = 0\}$. It can be easily seen that $\tilde{A} = A'/N$ is a Banach algebra under the norm $|||\cdot|||$. $\tilde{A}$ contains a subalgebra A_0 isometric to A, namely the subalgebra whose elements are represented by constant sequences. If $x_0 x_n \to 0$, $\|x_n\| = 1$, $x_0 \neq 0$, then the sequences (x_0), (x_n) are representatives of elements in $\tilde{A}$ which are divisors of zero there; and the coset containing the sequence (x_0) is in A_0.

Let A be a Banach algebra. Consider the operator $l_x \colon y \to xy$. The following situations can occur:

1° The operator l_x maps A onto itself in a one-to-one way. This happens if and only if x is invertible.

2° The operator l_x is not one-to-one. This happens if and only if x is a divisor of zero in A.

3° The operator l_x is one-to-one, the image $l_x A$ is closed in A but is not all of A. This happens if and only if x is not invertible in A and is not a topological divisor of zero.

4° The image $l_x A$ is not closed in A, then x is a topological divisor of zero.

Putting 3° and 4° together we obtain:

14.9. THEOREM. *An element x of a Banach algebra A, which is not a divisor of zero, is a topological divisor of zero if and only if $xA \neq \overline{xA}$.*

14.9.a. EXERCISE. Prove 14.9 in detail.

14.9.b. EXERCISE. Show by an example that $xA = \overline{xA}$ can happen for some divisor of zero, x. The assumption in Theorem 14.9 is thus essential.

14.9.c. EXERCISE. Prove that every non-zero element of the radical of a Banach algebra is a topological divisor of zero.

§ 15. THE SHILOV BOUNDARY

15.1. DEFINITION. A *maximizing set* for a Banach algebra A is any closed subset $F \subset \mathfrak{M}(A)$ such that

$$\sup_{\mathfrak{M}} |x\hat{\ }(M)| = \sup_{F} |x\hat{\ }(M)|$$

for each $x \in A$. A *Shilov boundary of A*, in symbols $\Gamma(A)$, is any minimal maximizing set.

15.2. THEOREM. *Every Banach algebra A has a unique Shilov boundary.*

PROOF. Let $\mathscr{F}$ denote the family of all maximizing sets for A. $\mathscr{F}$ is non-void since $\mathfrak{M} \in \mathscr{F}$. The family $\mathscr{F}$ is ordered by the inclusion. If (F_α) is a chain in $\mathscr{F}$, then the intersection $F = \bigcap F_\alpha$ is non-void by the compactness of F_α's. We shall show that $F \in \mathscr{F}$. In fact, for any $x \in A$ the set $S_x = \{M \in \mathfrak{M}: |x\hat{\ }(M)| = ||x||_s\}$ is compact and $F_\alpha \cap S_x \neq \emptyset$ for all α. Thus $F \cap S_x \neq \emptyset$ whence $F \in \mathscr{F}$. In virtue of the Kuratowski–Zorn lemma $\mathscr{F}$ has minimal elements. Now we show that there exists exactly one such minimal set. Suppose, on the contrary, that there are two Shilov boundaries Γ_1 and Γ_2. Let $M_0 \in \Gamma_1$ and let U be a neighbourhood of M_0: $U = \{M \in \mathfrak{M}: |\hat{x}_1(M)| < \varepsilon, \ldots, |\hat{x}_n(M)| < \varepsilon\}$, $x_1, \ldots, x_n \in M_0$. By the minimality of Γ_1 there exists an element $y \in A$ such that the Gelfand transform $y\hat{\ }(M)$ attains its maximum modulus $||y||_s$ on $U \cap \Gamma_1$ and $\max_{\Gamma_1 \setminus U} |y\hat{\ }(M)| < ||y||_s$, since otherwise the set $\Gamma_1 \setminus U$ would be a maximizing set smaller than Γ_1. We may assume that this maximal value is 1 and that

$$|y\hat{\ }(M)| < \frac{\varepsilon}{\max_{i \leqslant n} ||x_i||} \qquad \text{for } M \in \Gamma_1 \setminus U$$

replacing y, if necessary, by a suitable power y^m. So we have

$$\max_{\Gamma_1} |y\hat{}(M) x_i\hat{}(M)| = \max_{\mathfrak{M}} |y\hat{}(M) x_i\hat{}(M)| < \varepsilon.$$

Further, there exists an element $M_1 \in \Gamma_2$ such that $|y\hat{}(M_1)| = 1$ and, since $|x_i\hat{}(M_1) y\hat{}(M_1)| < \varepsilon$, so $|x_i\hat{}(M_1)| < \varepsilon$, $i = 1, 2, \ldots, n$. Thus $M_1 \in U$. The neighbourhoods (10.1.1) constitute a local base at M_0, so that every neighbourhood of M_0 contains points of $\Gamma_2 = \overline{\Gamma}_2$ and consequently $M_0 \in \Gamma_2$. It hence follows that $\Gamma_1 \subset \Gamma_2$ and, by the minimality of Γ_2, $\Gamma_1 = \Gamma_2$.

15.3. THEOREM. *A point $M_0 \in \mathfrak{M}(A)$ is in $\Gamma(A)$ if and only if for every neighbourhood U of M_0 a function $y\hat{} \in A\hat{}$ exists such that*

$$\sup_{\mathfrak{M}\setminus U} |y\hat{}(M)| < \sup_{U} |y\hat{}(M)|.$$

PROOF. If $M_0 \in \Gamma(A)$, then such a y exists, as shown in the proof of the preceding theorem. Conversely, if M_0 has the asserted property, then $\Gamma \cap U \neq \emptyset$ for any neighbourhood U of M_0, whence $M_0 \in \Gamma(A)$.

15.4. EXERCISES.

15.4.a. Find $\Gamma(A)$ for $A = C(\Omega)$.

15.4.b. Find $\Gamma(A)$ for $A = \mathscr{A}$ (Example 3.9).

15.4.c. Prove that if $A = [x_0]$ (cf. 10.8), then after the identification $\mathfrak{M}(A) = \sigma(x_0)$ (see 10.12) we have $\Gamma(A) = \partial\sigma(x_0)$, where ∂ denotes the topological boundary of a subset of the complex plane.

15.4.d. Let A be the algebra of all functions of two complex variables which are holomorphic in the bicylinder $B = \{(\xi, \eta) \in C^2 : |\xi| < 1, |\eta| < 1\}$ and continuous on its boundary. It is a Banach algebra under pointwise operations and the sup norm; $\mathfrak{M}(A) = B$. Prove that $\Gamma(A) = \{(\xi, \eta) \in C^2 : |\xi| = 1, |\eta| = 1\}$ and thus $\Gamma(A) \neq \partial\, \mathfrak{M}(A)$.

We shall investigate further properties of the Shilov boundary $\Gamma = \Gamma(A)$.

15.5. THEOREM. *For every $x \in A \in$ **BC**$_e$ we have*

(15.5.1) $$\partial[\sigma(x)] \subset x\hat{}(\Gamma).$$

PROOF. It follows from the compactness of Γ and the continuity of $x\hat{}$ that $x\hat{}(\Gamma)$ is a compact subset of the complex plane C. Thus if $\lambda_0 \notin x\hat{}(\Gamma)$, then

$$\min_{M \in \Gamma} |x\hat{}(M) - \lambda_0| > \delta$$

holds for some $\delta > 0$. If we had $\lambda_0 \in \partial[\sigma(x)]$, contrary to the assertion, then there would exist a $\lambda_1 \in \sigma(x)$ such that $|\lambda_0 - \lambda_1| < \frac{1}{2}\delta$. Then for any $M \in \Gamma$

$$|x\hat{}(M) - \lambda_1| \geqslant |x\hat{}(M) - \lambda_0| - |\lambda_0 - \lambda_1| > \delta - \tfrac{1}{2}\delta = \tfrac{1}{2}\delta,$$

so that for $u = (x - \lambda_1 e)^{-1}$ we have

$$||u||_s = \max_{M \in \Gamma} \left| \frac{1}{x\hat{}(M) - \lambda_1} \right| = (\min_{M \in \Gamma} |x\hat{}(M) - \lambda_1|)^{-1} < 2\delta^{-1}.$$

However, since $x\hat{}(M_0) = \lambda_0$ for some M_0, then

$$||u||_s \geqslant |u\hat{}(M)| = |\lambda_0 - \lambda_1|^{-1} > 2\delta^{-1}.$$

This contradiction proves formula (15.5.1).

Now we shall prove a theorem in a sense analogous to the classical Rouché theorem on analytic functions.

15.6. THEOREM. *Let A be a Banach algebra, $x, y \in A$, and*

$$|x\hat{\ }(M)-y\hat{\ }(M)| < |x\hat{\ }(M)| \tag{15.6.1}$$

for all $M \in \Gamma(A)$. Then the element x is singular if and only if the element y is singular.

PROOF. It follows by the compactness of $\Gamma = \Gamma(A)$ and inequality (15.6.1) that there exists an integer $n > 1$ such that

$$n \min(|x\hat{\ }(M)|-|x\hat{\ }(M)-y\hat{\ }(M)|) > \|x-y\|_s. \tag{15.6.2}$$

Consider the finite sequence

$$nx,\ (n-1)x+y,\ (n-2)x+2y,\ \ldots,\ (n-k)x+ky,\ \ldots,\ ny.$$

If the assertion is not true, then either the first or the last term is invertible whereas the other one is not. Thus there exists an invertible term $(n-k)x+ky$ which follows or is followed by a singular one, i.e. by one belonging to a maximal ideal $M_0 \in \mathfrak{M}$. Write $u = [(n-k)x+ky]^{-1}$. We have $\max_{\Gamma} |u\hat{\ }(M)| = \max_{\mathfrak{M}} |u\hat{\ }(M)|$ so that $\min_{\Gamma} |u\hat{\ }(M)^{-1}| = \min_{\mathfrak{M}} |u\hat{\ }(M)^{-1}|$. Hence and by (15.6.2) follows the estimate:

$$\begin{aligned}
\|x-y\|_s &< n \min_{\Gamma} [|x\hat{\ }(M)|-|x\hat{\ }(M)-y\hat{\ }(M)|] \\
&\leqslant \min_{\Gamma} [n|x\hat{\ }(M)|-k|x\hat{\ }(M)-y\hat{\ }(M)|] \\
&\leqslant \min_{\Gamma} |nx\hat{\ }(M)-k[x\hat{\ }(M)-y\hat{\ }(M)]| \\
&= \min_{\mathfrak{M}} |nx\hat{\ }(M)-k[x\hat{\ }(M)-y\hat{\ }(M)]| \\
&\leqslant |nx\hat{\ }(M_0)-k[x\hat{\ }(M_0)-y\hat{\ }(M_0)]|.
\end{aligned}$$

Since the element $(n-k\pm 1)x+(k\mp 1)y$ is in M_0, then

$$\begin{aligned}
\|x-y\|_s &< |nx\hat{\ }(M_0)-k[x\hat{\ }(M_0)-y\hat{\ }(M_0)]| \\
&= |nx\hat{\ }(M_0)-k[x\hat{\ }(M_0)-y\hat{\ }(M_0)]- \\
&\quad -[(n-k\pm 1)x\hat{\ }(M_0)+(k\mp 1)y\hat{\ }(M_0)]| \\
&= |x\hat{\ }(M_0)-y\hat{\ }(M_0)| \leqslant \|x-y\|_s
\end{aligned}$$

and the obtained contradiction concludes the proof.

We now show how the notion of the Shilov boundary can be applied to the formulation of a theorem on the extension of multiplicative linear functionals.

15.7. THEOREM. *Let A_0 be a subalgebra of a Banach algebra A. Then every ideal $M \in \Gamma(A_0)$ is contained in a maximal ideal belonging to $\mathfrak{M}(A)$. In other words, every multiplicative linear functional on A_0 belonging to $\Gamma(A_0)$ admits an extension to a multiplicative linear functional on all of A.*

PROOF. Let $x \in A_0 \subset A$. Since the value of the spectral norm $\|x\|_s = \lim \sqrt[n]{\|x^n\|}$ does not depend on the algebra containing x, then

$$\|x\|_s = \max_{\Gamma(A_0)} |x\hat{}(M)| = \max_{\mathfrak{M}(A)} |x\hat{}(M)|.$$

If an ideal $M_0 \in \Gamma(A_0)$ does not extend to a proper ideal in A, then the sums $\sum_{i=1}^{n} x_i z_i$, $x_i \in M_0 \subset A_0$, $z_i \in A$, $i = 1, 2, \ldots, n$, exhaust the whole algebra A; in particular, then $e = x_1 z_1 + \ldots + x_n z_n$ for some x_i, z_i, $i = 1, 2, \ldots, n$. Without loss of generality we may assume that $\|x_i\|_s \leqslant 1$. Let $\mu > \max(\|z_1\|_s, \ldots, \|z_n\|_s)$. Consider the neighbourhood U of M_0 in $\mathfrak{M}(A_0)$ defined by the inequalities $|x_i\hat{}(M)| < 1/2n\mu$. Let y be an element of A_0 such that $\sup_U |y\hat{}(M)| = 1$ and $\sup_{\mathfrak{M}(A_0)\setminus U} |y\hat{}(M)| < 1/2n\mu$ (such a y exists by Theorem 15.3). Thus we have

$$(15.7.1) \qquad 1 = \|y\|_s = \Big\| y \sum_{i=1}^{n} x_i z_i \Big\|_s \leqslant \sum_{i=1}^{n} \|y x_i\|_s \|z_i\|_s$$
$$\leqslant \mu \sum_{i=1}^{n} \|x_i y\|_s = \mu \sum_{i=1}^{n} \max_{\Gamma(A_0)} |x_i\hat{}(M) y\hat{}(M)|.$$

But

$$\sup_U |x_i\hat{}(M) y\hat{}(M)| \leqslant \sup_U |x_i\hat{}(M)| \sup_U |y\hat{}(M)| \leqslant 1/2n\mu.$$

Also for $M \in \Gamma(A_0)\setminus U$ we have

$$\max_{\Gamma(A_0)\setminus U} |x_i\hat{}(M) y\hat{}(M)| \leqslant \max_{\Gamma\setminus U} |x_i\hat{}(M)| \max_{\Gamma\setminus U} |y\hat{}(M)| = 1/2n\mu;$$

thus the sum on the right-hand side of (15.7.1) does not exceed $\frac{1}{2}$ which gives a contradiction concluding the proof.

15.7.a. EXERCISE. Let $A_0 = \mathscr{A}$ (Example 3.9), $A_0 \subset A = C(\{z \in C: |z| = 1\})$. Show that the only multiplicative linear functionals on A_0 which extend to functionals on A are the elements of the Shilov boundary $\Gamma(A_0)$.

It can happen that the set of those maximal ideals of A which are extendible to ideals in any superalgebra of A is actually larger than the Shilov boundary $\Gamma(A)$. In connection therewith we give the following definition:

15.8. DEFINITION. Let A be a Banach algebra. The *cortex of* A, in symbols $\operatorname{cor} A$, is the set of those elements $f \in \mathfrak{M}(A)$ which admit an extension to multiplicative linear functionals on any superalgebra of A.

15.9. THEOREM. *Let A be a Banach algebra. Then we have $\Gamma(A) \subset \operatorname{cor} A$ and $\operatorname{cor} A$ is a compact subset of $\mathfrak{M}(A)$.*

PROOF. The inclusion $\Gamma(A) \subset \operatorname{cor} A$ is asserted by the preceding theorem. Let us consider an arbitrary extension B of A and denote by φ_B the restriction map

$\mathfrak{M}(B) \to \mathfrak{M}(A)$ given by $f \overset{\varphi_B}{\to} f|_A$. φ_B maps the multiplicative linear functionals of B into the multiplicative linear functionals of A. The map φ_B is clearly continuous, thus $\varphi_B(\mathfrak{M}(B))$ is a compact subset of $\mathfrak{M}(A)$. The compactness of the cortex of A follows now from the relation

$$\operatorname{cor} A = \bigcap \varphi_B(\mathfrak{M}(B)),$$

the intersection of all subsets of $\mathfrak{M}(A)$ which are of the form $\varphi_B(\mathfrak{M}(B))$, where B is some extension of A.

15.10. DEFINITION. A subset S of a Banach algebra A consists of *joint topological divisors of zero* if

(15.10.1)
$$\inf_{\|z\|=1,\, z\in A} \sum_{i=1}^{n} \|zx_i\| = 0$$

for every finite subset $\{x_1, \ldots, x_n\} \subset S$.

15.10.a. EXERCISE. Prove that if a maximal ideal M of A consists of joint topological divisors of zero, then $M \in \operatorname{cor} A$.

In view of the above exercise we give the following:

15.11. CONJECTURE. Let $A \in \mathbf{BC}_e$ and $M \in \mathfrak{M}(A)$. Then $M \in \operatorname{cor} A$ if and only if it consists of joint topological divisors of zero.

Another characterization of the cortex is given in Arens [9]. The conjecture 15.11 can be supported by the following result (Żelazko [15]).

15.12. THEOREM. *If $A \in \mathbf{BC}_e$ and $M \in \Gamma(A)$, then M consists of joint topological divisors of zero* (cf. also Żelazko [13]).

COMMENTS ON CHAPTER II

C.8. All given here results are true for commutative p-normed algebras (cf. Żelazko [8]). As can be seen from Exercise 8.4.a, Theorems 8.2 and 8.4 are not true in the algebra $\mathscr{E}$ (Example C.3.3) and so it fails even for m-convex B_0-algebras. It is still an open question whether Theorem 8.3 is true for m-convex B_0-algebras. (The problem for m-convex can be reduced to certain algebras consisting of analytic functions, cf. Craw [2]. In Dixon and Fremlin [1] it is shown that this problem is equivalent to the question whether multiplicative linear functionals in m-convex algebras are bounded on bounded sets.) Some information is given by

C.8.1. THEOREM. *Let A be a commutative m-convex B_0-algebra with unit. If A has a finite number of generators, then every multiplicative linear functional on A is continuous* (cf. Arens [5]).

In non-metrizable complete m-convex algebras one may have discontinuous multiplicative linear functionals. It is so e.g. in the algebra $C(T)$—with compact-open topology, where T is the set of all ordinals smaller than Ω — the first uncountable ordinal, with the order topology (cf. Michael [1]). Corollary 8.7 is true for m-convex algebras, moreover, in any such algebra with unit there exists at least one continuous multiplicative and linear functional (this follows immediately from Theorem C.2.3).

For non-m-convex B_0-algebras this result fails, one can show that in L^ω (Example C.3.4) there are no non-zero multiplicative linear functionals (cf. Żelazko [8]).

C.8.2. NOTATION. If $A \in \mathbf{CM}_e$ (**CM** = commutative m-convex), then we write $\mathfrak{M}(A)$ for the set of all continuous non-zero multiplicative linear functionals on A, and denote by $\mathfrak{M}^{\#}(A)$ the set of all non-zero multiplicative linear functionals on A, so that $\mathfrak{M}(A) \subset \mathfrak{M}^{\#}(A)$.

With this notation we have an analogue of Theorem 8.11:

C.8.3. THEOREM. *Let $A \in \mathbf{CM}_e$; then an element $x \in A$ is invertible if and only if $f(x) \neq 0$ for all $f \in \mathfrak{M}(A)$.*

This theorem is an immediate consequence of Theorem C.2.3 (cf. also Michael [1]).

For m-convex algebras we shall use in the sequel the notation $x^\wedge$ meaning by that the map $f \to f(x)$ defined on $\mathfrak{M}(A)$ or $\mathfrak{M}^{\#}(A)$.

C.9. Formula (9.4.2) gives also the general form of a multiplicative linear functional in the algebra l_p (Example 6.3.1). So, as in Exercise 9.4.b we can obtain the following generalization of the Wiener theorem (cf. Żelazko [8]).

C.9.1. THEOREM. *Let x be a complex function of real variable t equal to its Fourier expansion*

$$x(t) = \sum_{-\infty}^{\infty} x_n e^{int}$$

and suppose that

$$\sum |x_n|^p < \infty,$$

where $0 < p \leqslant 1$. If $x(t) \neq 0$ for all t, then its inverse has Fourier expansion

$$\frac{1}{x(t)} = \sum y_n e^{int}$$

with

$$\sum |y_n|^p < \infty.$$

C.10. All results presented here are also valid for p-normed algebras. If A is a commutative m-convex algebra with unit, then the Gelfand topology can be introduced in $\mathfrak{M}(A)$ or even in $\mathfrak{M}^{\#}(A)$ (cf. Notation C.8.2), but, obviously, $\mathfrak{M}(A)$ need not be a compact space in this case, and it even need not be a locally compact space.

In the case when A is a commutative m-convex B_0-algebra, then $\mathfrak{M}(A)$ provided with the Gelfand topology is a so-called hemicompact space, i.e. a topological space for which there exists a sequence $(K_n)_0^\infty$ of compact subsets whose union is the whole space and such that every compact set in $\mathfrak{M}(A)$ is contained in some K_i.

The definition of a spectrum or a joint spectrum for an m-convex algebra is the same as for a Banach algebra, but the spectrum need not be compact. It is, however, never void, and can be given by the formula

$$\sigma_A(x) = \bigcup_{\alpha} \sigma_{A_\alpha}(\pi_\alpha(x)) = x^\wedge(\mathfrak{M}(A)) = x^\wedge(\mathfrak{M}^{\#}(A))$$

(cf. notation of C.2 and notation C.8.2).

The problem whether for finitely generated commutative m-convex algebras the space $\mathfrak{M}(A)$ is homeomorphic with the joint spectrum $\sigma_A(x_1, \ldots, x_n)$ of the generators has been discussed in Brooks [4], where it is shown that

(i) It may occur that one generating family $(x_1, \ldots, x_n) \subset A$ yields a topological map (of $\mathfrak{M}(A)$ onto $\sigma(x_1, \ldots, x_n)$), while another fails to.

(ii) It may be that no generating family induces a homeomorphism.

(iii) The joint spectrum need not be polynomially convex.

For examples realizing (i) and (ii) we can take singly generated m-convex B_0-algebras

Concerning general LC-algebras the situation is more complicated, since by results mentioned in C.5 the spectrum of an element $x \in A \in \mathrm{LC}$ can be empty even if A is a B_0-algebra (it is so in the case when A contains all rational functions evaluated at x). There are several, essentially equivalent, approaches for generalizing the concept of spectrum for an LC-algebra. We describe one of them (given in Żelazko [8], for others see Waellbroeck [1], Neubauer [1], [2], Allan [1]). We give namely the following

C.10.1. DEFINITION. Let A be an LC-algebra with unit and $G(A)$ the group of all its invertible elements. We put

(C.10.1) $$\sigma(x) = \{\lambda \in C: x-\lambda e \in G(A)\},$$

(C.10.2) $$\sigma_d(x) = \{\lambda_0 \in C\setminus\sigma(x): \text{the map } \lambda \to (x-\lambda e)^{-1} \text{ is discontinuous at } \lambda_0\},$$

(C.10.3) $$\sigma_\infty(x) = \begin{cases} \emptyset & \text{if the map } \lambda \to (x-\lambda e)^{-1} \text{ is continuous at } \lambda = 0, \\ \infty & \text{otherwise} \end{cases}$$

and finally

$$\Sigma_A(x) = \sigma(x) \cup \sigma_d(x) \cup \sigma_\infty(x).$$

The set $\Sigma_A(x)$ is called the *extended spectrum of an element* $x \in A$, it is a subset of the closed complex plane (Riemann sphere) and can contain the point at infinity.

We mention now some results on the extended spectrum justifying the introduction of this concept (if A has no unit, then as in remarks after Theorem 17.1 we first embed it in $A_1 = A \oplus \{\lambda e\}$ and then define the spectrum $\Sigma_A(x)$ as that in A_1).

C.10.2. THEOREM. *If $A \in LC$, then $\Sigma_A(x)$ is never void (but each of the sets $\sigma(x) \cup \sigma_d(x)$, $\sigma_d(x) \cup \sigma_\infty(x)$, $\sigma(x) \cup \sigma_\infty(x)$ can be void) and has the following properties*

(i) *If $x \in G(A)$, then*

$$\Sigma_A(x^{-1}) = (\Sigma_A(x))^{-1}$$

(it is possible to have $0 \in \Sigma_A(x)$ for an invertible x, so we put here $\infty^{-1} = 0$ and $0^{-1} = \infty$).

(ii) *If $\lambda \in \Sigma_A(x)$, then $\lambda^2 \in \Sigma_A(x^2)$.*

Let us mention also here, that the concept of spectrum can be introduced and is quite natural in more general objects then topological algebras, namely in algebras with an axiomatically defined family of "bounded sets" (algebras with a boundedness). The spectral theory in such algebras has been extensively studied by L. Waelbroeck (cf. [3]–[7]).

C.12. All results given here are true for p-normed algebras, so for a p-normed algebra the unit ball of its spectral norm is a convex set (upon this fact one can build up a substantial part of the theory of these algebras).

For complete m-convex algebras the results are also true (to each seminorm $||x||_\alpha$ there corresponds a suitable spectral seminorm given by $\lim \sqrt[n]{||x^n||_\alpha}$).

Define the spectral radius for $x \in A \in \mathrm{LC}$ as

$$\varrho(x) = \sup\{|\lambda|: \lambda \in \Sigma_A(x)\}.$$

If $\infty \in \Sigma_A(x)$ we put $\varrho(x) = \infty$.

Under this notation we have the following

C.12.1. THEOREM. *For $x \in A \in LC$ put*

$$r_1(x) = \sup_\alpha \limsup \sqrt[n]{||x^n||_\alpha}, \qquad r_2(x) = \sup_{f \in A} \limsup \sqrt[n]{|f(x^n)|},$$

$$r_3(x) = \inf\{r > 0: \text{for every power series } \textstyle\sum a_n t^n \text{ with complex coefficients } a_n \text{ and with radius of convergence} \geqslant r \text{ the series } \textstyle\sum a_n x^n \text{ is convergent in } A\}.$$

Then $r_1(x) = r_2(x) = r_3(x) = \varrho(x)$ (cf. Żelazko [8]).

The radical in an m-convex algebra is defined as the intersection of all closed maximal ideals (= kernels of elements in $\mathfrak{M}(A)$) and Theorem 12.9 is satisfied.

For non-m-convex algebras, however, the set $\mathfrak{M}(A)$ can be void (cf. C.8), and so in this case the definition of radical should rather be given by

$$\operatorname{rad} A = \{x \in A: e+xy \in G(A) \text{ for all } y \in A\}.$$

The trouble is that in a non-m-convex B_0-algebra the radical, which is always a proper ideal in an algebra with unit, can be a non-closed ideal (for an example cf. Rolewicz [4]). Another trouble is that if $\operatorname{rad} A$ is non-closed, then the quotient $A/\overline{\operatorname{rad} A}$ can be a non-semisimple algebra (cf. Dixon [3]).

It can be also shown that if A is a commutative B_0-algebra and if its radical $\operatorname{rad} A$ is closed in A, then $\operatorname{rad} A$ is an m-convex algebra (cf. Żelazko [8]).

In a commutative m-convex algebra A the radical is characterized by: $x \in \operatorname{rad} A$ if and only if $\lim \sqrt[n]{\|x^n\|_\alpha} = 0$ for every continuous seminorm $\|x\|_\alpha$ on A, or, by Theorem C.12.1, if $\varrho(x) = 0$. This implies that if A_0 is a subalgebra of an m-convex algebra A, and if $x \in \operatorname{rad} A_0$, then $x \in \operatorname{rad} A$. This is no longer true in the non-m-convex case and there exists a non-m-convex B_0-algebra A having an m-convex subalgebra A_0 with non-zero radical, such that all non-zero elements of $\operatorname{rad} A_0$ are invertible in A (cf. Żelazko [14]).

This implies also that the elements of a radical need not be topological divisors of zero, even in the case of an m-convex B_0-algebra, while for Banach algebras it must be, so we mention two results about escaping of an element from a radical after taking a suitable extension.

C.12.2. THEOREM. *Let $A \in BC_e$ and let $x \in \operatorname{rad} A$. Then there is a B_0-algebra B which is a superalgebra for A (and containing A isometrically) such that $x \in \operatorname{rad} B$ if and only if x is not a nilpotent element* (cf. Rolewicz [4]).

In order to have a generalization of this result to B_0-algebras we need the following definition.

C.12.3. DEFINITION. Let A be a B_0-algebra; $x \in A$ is called *almost nilpotent* if for any continuous seminorm $\|x\|$ in A, there is a natural number n such that $\|x^n\| = 0$ (or, equivalently, if $\|x^n\|_i = 0$ for $n \geqslant n(i)$ for $i = 1, 2, \ldots$).

Clearly, any almost nilpotent element is in the radical $\operatorname{rad} A$.

There are almost nilpotent elements which are not nilpotent (cf. Kitainik [1]).

C.12.4. THEOREM. *Let A be a commutative B_0-algebra and let $x \in \operatorname{rad} A$. Then there is a B_0-algebra B being a superalgebra for A (and containing A isometrically) such that $x \in \operatorname{rad} B$ if and only if x is not an almost nilpotent element* (cf. Kitainik [1]).

C.14. All results of this section are true also for p-normed algebras. Theorem 14.2 fails even for m-convex B_0-algebras, e.g. the algebra $\mathscr{E}$ (Example C.3.3) possesses no topological divisors of zero. The concept, however, can be generalized as follows:

C.14.1. DEFINITION. Let A be a topological algebra. Call two non-void subsets $U, V \subset A$ *generalized topological divisors of zero*, if $0 \notin \overline{U} \cup \overline{V}$ but $0 \in \overline{UV}$.

With this definition we have the following

C.14.2. THEOREM. *Let A be an m-convex algebra. Then either A is isomorphic to the field of complex numbers, or A possesses generalized topological divisors of zero.*

A similar result is true for real m-convex algebras, of course in the conclusion instead of merely complex numbers we can have also the field of real numbers or the division algebra of quaternions (for proofs cf. Żelazko [9] and [10]).

We conjecture that Theorem C.14.2 is true for all topological algebras. As a justification for this conjecture we quote the following result.

C.14.3. Let A be a topological algebra with unit, and suppose that for every $x \in A$, $x \neq 0$, there is an inverse $x^{-1} \in A$. Then either A possesses generalized topological divisors of zero, or $A = \boldsymbol{C}$ (if A is over the complex field), or $A = \boldsymbol{R}$ or $\boldsymbol{C}$ or $\boldsymbol{Q}$ (if A is over real scalars) (cf. Żelazko [8]).

This justifies our conjecture since division algebras should be as far as possible from having divisors of zero of any kind.

Concerning permanently singular elements in m-convex algebras we give here the following definition

C.14.4. Definition. Let k be a class of topological algebras. We call an element $x \in A \in k$ *k-singular*, if for any superalgebra $B \supset A$ (i.e. topological algebra containing A isomorphically) of class k the element x is non-invertible in B.

With this notation we have the following result.

C.14.5. Theorem. *Let $A \in M$ and $x \in A$. Then x is M-singular if and only if for any system $(||x||_\alpha)$ of submultiplicative semi-norms defining the topology of A there is an index α such that $\pi_\alpha(x)$ is a topological divisor of zero in A_α* (cf. Notation of C.2.).

What is more interesting, is that one can construct a non-m-convex B_0-algebra A and its m-convex subalgebra A_0 in such a way that there are in A_0 M-singular elements which are invertible in A (cf. Żelazko [14]).

This implies that even in the m-convex case the permanently singular elements need not be topological divisors of zero and that the concept of permanent singularity depends upon the considered class (this is also a justification for Definition C.14.4), while for Banach algebras an element which is singular with respect to the class B is also singular with respect to all topological algebras.

C.15. All results given here hold also for p-normed algebras. There are no satisfactory attempts to study similar concepts for LC-algebras. For certain m-convex function algebras a concept of boundary was introduced and studied in Rickart [2], and Brooks [3] and [6].

Let us mention that there is also no satisfactory characterization of the cortex (Definition 15.8) even for Banach algebras (cf. conjecture 15.11).

CHAPTER III

Analytic Functions in Banach Algebras

§ 16. ANALYTIC FUNCTIONS OF ONE COMPLEX VARIABLE

16.1. DEFINITION. Let $A \in \mathbf{BC}_e$ and let φ be a complex-valued function of a complex variable defined on a subset $D \subset \boldsymbol{C}$. The function φ is said *to act on an element* $x \in A$ if $\sigma(x) \subset D$ and $\varphi[x^{\wedge}(\cdot)] \in A^{\wedge}$. For instance, any entire function $\varphi(\lambda) = \sum_{n=0}^{\infty} a_n \lambda^n$ acts on each element of any Banach algebra, since by a formal substitution of x in place of λ we obtain the series $\sum_{n=0}^{\infty} a_n x^n$ which is clearly convergent in A to an element y. We then have $y^{\wedge}(M) = \sum_{n=0}^{\infty} a_n [x^{\wedge}(M)]^n = \varphi[x^{\wedge}(M)]$ so that the function φ acts on x.

We now prove the following theorem.

16.2. THEOREM. *Let $A \in \mathbf{BC}_e$ and let φ be a function defined and holomorphic in an open set $\Omega \subset \boldsymbol{C}$. Then φ acts on every element $x \in A$ such that $\sigma(x) \subset \Omega$.*

In other words, *for every such element $x \in A$ there exists a $y \in A$ such that*

$$f(y) = \varphi[f(x)] \tag{16.2.1}$$

holds for any functional $f \in \mathfrak{M}(A)$.

The proof will be preceded by a number of remarks concerning continuous functions with values in a Banach space.

Let X be a Banach space with a norm $\|\cdot\|$, let γ be a compact oriented rectifiable arc in the complex plane $\boldsymbol{C}$ and let $x(\lambda)$ be a continuous X-valued function defined on γ. It can be proved, similarly as is done with complex-valued functions, that the function $x(\lambda)$ is uniformly continuous and that the limit

$$\lim_{\max_k |\lambda_{k+1} - \lambda_k| \to 0} \sum_{k=0}^{n-1} x(\lambda_k') \, (\lambda_{k+1} - \lambda_k)$$

exists, where $\lambda_i, \lambda_i' \in \gamma$ and each λ_k' lies between λ_k and λ_{k+1}; The ordering of λ_k's is assumed to coincide with the orientation of γ. This limit will be denoted by the symbol $\int_\gamma x(\lambda)\,d\lambda$; it is an element of X. It follows directly from this definition of the integral that the equality

$$f\left[\int_\gamma x(\lambda)\,d\lambda\right] = \int_\gamma f[x(\lambda)]\,d\lambda$$

holds for any functional $f \in X'$, where the integral occurring on the right-hand side is the usual integral of a complex-valued function.

We now pass to the proof of Theorem 16.2.

Let x be a fixed element of A such that $\sigma(x) \subset \Omega$. Since the spectrum $\sigma(x)$ is compact and is covered by the components of the open set Ω, it is also covered by finitely many components. We may thus assume that Ω has finitely many components. The distance between the sets $\sigma(x)$ and $\mathbf{C} \setminus \Omega$ is positive, consequently there exists a finite system of compact rectifiable arcs in Ω, $\gamma_1, \ldots, \gamma_n$, whose union is the boundary of an open set Ω_0 with $\sigma(x) \subset \Omega_0 \subset \Omega$. The values of φ in Ω_0 are given by the Cauchy formula

$$\varphi(\xi) = \sum_{k=1}^{n} \frac{1}{2\pi i} \int_{\gamma_k} \frac{\varphi(\lambda)}{\lambda - \xi}\,d\lambda.$$

Substituting the element x formally in place of ξ we obtain an element y of A:

$$y = \sum_{k=1}^{n} \frac{1}{2\pi i} \int_{\gamma_k} \varphi(\lambda)\,(\lambda e - x)^{-1}\,d\lambda. \tag{16.2.2}$$

This integral is defined since the function $x(\lambda) = \varphi(\lambda)\,(\lambda e - x)^{-1}$ is continuous along each arc γ_k. Further,

$$f(y) = \sum_{k=1}^{n} \frac{1}{2\pi i} \int_{\gamma_k} f[\varphi(\lambda)\,(\lambda e - x)^{-1}]\,d\lambda = \sum_{k=1}^{n} \frac{1}{2\pi i} \int_{\gamma_k} \frac{\varphi(\lambda)}{\lambda - f(x)}\,d\lambda = \varphi[f(x)]$$

for all $f \in \mathfrak{M}$, since the curves γ_k encompass the spectrum $\sigma(x)$. Consequently relation (16.2.1) holds.

16.2.a. Exercise. Show that the element y defined by (16.2.2) does not depend on the choice of the curves γ_k provided they bound an open set containing the spectrum of x, and that y is uniquely determined by φ.

16.2.b. Exercise. Show that an element y satisfying relation (16.2.1) is uniquely determined by x if and only if A is semi-simple.

We are now going to prove that the element y assigned to x and φ is uniquely determined by the additional requirement that this assignment should be a homomorphism when x is fixed and φ ranges over holomorphic functions. More precisely:

16.3 Definition. Let D be a compact subset of the complex plane $\mathbf{C}$. Let A_D denote the algebra of all functions holomorphic in a neighbourhood of D (depending on the function). A sequence $(\varphi_n) \subset A_D$ is said *to converge* if all functions φ_n are holomorphic in some common neighbourhood of D and the sequence φ_n converges uniformly in this neighbourhood. The space A_D is clearly complete in the sense of this convergence.

16.4. THEOREM. *Let $A \in \mathbf{BC}_e$ and let D be a compact subset of the complex plane C. Let x be a fixed element of A whose spectrum is contained in D. Then there exists exactly one mapping $F_x: \varphi \to y = \varphi(x)$ of A_D into A satisfying the following conditions:*

(i) *If $\varphi(\lambda) \equiv \lambda$, then $\varphi(x) = x$.*

(ii) *If $\varphi(\lambda) \equiv 1$, then $\varphi(x) = e$.*

(iii) *If $\varphi(\lambda) = \alpha\varphi_1(\lambda)+\beta\varphi_2(\lambda)$, then $\varphi(x) = \alpha\varphi_1(x)+\beta\varphi_2(x)$, $\alpha, \beta \in C$.*

(iv) *If $\varphi(\lambda) = \varphi_1(\lambda)\varphi_2(\lambda)$, then $\varphi(x) = \varphi_1(x)\varphi_2(x)$.*

(v) *If $\varphi_n \to 0$ in A_D, then $\varphi_n(x) \to 0$ in A.*

This mapping is given by formula (16.2.2).

PROOF. We first show that the mapping $\varphi \to \varphi(x)$ given by (16.2.2) in fact satisfies conditions (i)–(v).

(i). The expansion

$$(\lambda e - x)^{-1} = \lambda^{-1}e+\lambda^{-2}x+\lambda^{-3}x^2+\ldots$$

holds for large $|\lambda|$ (e.g. for $|\lambda| > \|x\|$), thus the function $\varphi(\lambda) \equiv \lambda$ satisfies the equality

$$\varphi(x) = \frac{1}{2\pi i}\int\limits_{|\lambda|=R} \lambda(\lambda e-x)^{-1}d\lambda$$

$$= \frac{1}{2\pi i}\int\limits_{|\lambda|=R}\sum_{n=0}^{\infty}\lambda^{-n}x^n d\lambda = \frac{1}{2\pi i}\sum_{n=0}^{\infty}\int\limits_{|\lambda|=R}\frac{d\lambda}{\lambda^n}x^n = x$$

whenever $R > \max\{|\xi| : \xi \in D\}$.

(ii) is proved similarly. (iii) is obvious. (iv) will be proved if we show that

$$(16.4.1) \qquad \frac{1}{2\pi i}\int\limits_{\gamma}(\lambda e-x)^{-1}\varphi_1(\lambda)\varphi_2(\lambda)\,d\lambda$$

$$= \frac{1}{2\pi i}\int\limits_{\gamma}(\lambda e-x)^{-1}\varphi_1(\lambda)d\lambda\,\frac{1}{2\pi i}\int\limits_{\gamma}(\mu e-x)^{-1}\varphi_2(\mu)\,d\mu,$$

where γ is a system of closed curves encompassing the set D and contained in the intersection $V = U_1 \cap U_2$ of the sets of holomorphicity of the functions φ_1 and φ_2. Since the value of integral (16.2.2) is independent of the choice of the path of integration (Exercise 16.2.a), then the curve γ may be replaced in the last integral in (16.4.1) by a parallel larger curve γ_1 contained in V (larger in the sense that the Cauchy integral formula with integration performed along γ_1 defines a function holomorphic at each point of γ). After such a replacement the right-hand side of (16.4.1) is equal to

$$-\frac{1}{4\pi^2}\int_\gamma\int_{\gamma_1}(\lambda e-x)^{-1}(\mu e-x)^{-1}\varphi_1(\lambda)\varphi_2(\mu)\,d\lambda\,d\mu$$

$$=-\frac{1}{4\pi^2}\int_\gamma\int_{\gamma_1}\frac{\varphi_1(\lambda)\varphi_2(\mu)}{\mu-\lambda}[(\lambda e-x)^{-1}-(\mu e-x)^{-1}]\,d\lambda\,d\mu$$

$$=\frac{1}{2\pi i}\int_\gamma(\lambda e-x)^{-1}\varphi_1(\lambda)\left[\frac{1}{2\pi i}\int_{\gamma_1}\frac{\varphi_2(\mu)}{\mu-\lambda}\,d\mu\right]d\lambda+$$

$$+\frac{1}{4\pi^2}\int_{\gamma_1}(\mu e-x)^{-1}\varphi_2(\mu)\left[\int_\gamma\frac{\varphi_1(\lambda)}{\mu-\lambda}\,d\lambda\right]d\mu.$$

But $\dfrac{\varphi_1(\lambda)}{\mu-\lambda}$ is a holomorphic function of the variable λ for any fixed $\mu\in\gamma_1$, so that $\displaystyle\int_\gamma\frac{\varphi_1(\lambda)}{\mu-\lambda}\,d\lambda=0$, whence (16.4.1) follows.

In order to prove (v) it suffices to observe that

$$\|\varphi(x)\|\leqslant\frac{|\gamma|}{2\pi}\max_{\lambda\in\gamma}|\varphi(\lambda)|\max_{\lambda\in\gamma}\|(x-\lambda e)^{-1}\|,$$

where $|\lambda|$ denotes the sum of the lengths of all arcs in γ.

Now we shall prove that if a mapping $F_x\colon A_D\to A$ satisfies conditions (i)–(v), then it is given by $F_x(\varphi)=\varphi(x)$ with $\varphi(x)$ defined by formula (16.2.2).

Let $\varphi\in A_D$. If $\lambda_0\notin D$, then $(\lambda-\lambda_0)^{-1}\in A_D$ for $\lambda\in D$ and $F_x[(\lambda-\lambda_0)^{-1}]=(x-\lambda_0 e)^{-1}$ by (i), (ii), (iii), (iv). Given a function φ holomorphic in an open set U, we may write

$$\varphi(\lambda)=\frac{1}{2\pi i}\int_\gamma\frac{\varphi(\xi)}{\xi-\lambda}\,d\xi, \tag{16.4.2}$$

where again γ is a closed curve (a system of closed curves) contained in U and containing D in its interior. Integral (16.4.2) is precisely the limit of the following sequence

$$\varphi_n(\lambda)=\frac{1}{2\pi i}\sum_{k=0}^{k_n}\frac{\varphi(\xi'_{nk})\,(\xi_{n,k+1}-\xi_{nk})}{\xi'_{nk}-\lambda};$$

this sequence converges to φ uniformly on γ, thus also in some neighbourhood of D. Since

$$F_x(\varphi_n)=\frac{1}{2\pi i}\sum_{k=0}^{k_n}\varphi(\xi'_{nk})\,(\xi_{n,k+1}-\xi_{nk})\,(\xi'_{nk}e-x)^{-1},$$

then the sequence $F_x(\varphi_n)$ converges in A, by condition (v), and

$$\lim F_x(\varphi_n)=F_x(\varphi).$$

On the other hand,

$$\lim F_x(\varphi_n) = \frac{1}{2\pi i}\int_\gamma \varphi(\lambda)\,(\lambda e - x)^{-1} d\lambda = \varphi(x),$$

so that $F_x(\varphi) = \varphi(x)$.

The calculation of the values $\varphi(x)$ of holomorphic functions at points of A is called the *symbolic calculus* or the *operator calculus* in the theory of Banach algebras. The definition of $\varphi(x)$ as an image of φ under a certain homomorphism is better than that given by formula (16.2.1), since the latter is unique in the case of semi-simple algebras only and only makes sense for commutative algebras which possess multiplicative linear functionals. Note that in any (possibly non-commutative) Banach algebra any element x always commutes with the element $\varphi(x)$ which always can be constructed by means of the maximal commutative subalgebra containing x.

16.5. Remark. As it was recently shown by Allan [4], assumption (v) is essential in the proof of Theorem 16.4, so that in certain radical algebras there can exist other (discontinuous) maps satisfying conditions (i)–(iv).

16.6. Exercises.

16.6a. Prove the following *Levy theorem* (cf. Exercise 9.4.b):

If $x(t) = \sum_{n=-\infty}^{\infty} x_n e^{int}$, $0 \leqslant t < 2\pi$, $\sum_{-\infty}^{\infty} |x_n| < \infty$, *and if* φ *is a function holomorphic in a neighbourhood of the set of the values* $x(t)$, *then*

$$\varphi[x(t)] = \sum_{n=-\infty}^{\infty} y_n e^{int},$$

where $\sum_{n=-\infty}^{\infty} |y_n| < \infty$.

16.6.b. Prove that if a function continuous in the closed unit disc, holomorphic in its interior is given by a series $x(\lambda) = \sum_{n=0}^{\infty} a_n \lambda^n$ with $\sum_{n=0}^{\infty} |a_n| < \infty$ and if φ is a function holomorphic in a neighbourhood of the set of values of x, then the composition $\varphi(x(\lambda))$ also has an expansion $\varphi(x(\lambda)) = \sum_{n=0}^{\infty} b_n \lambda^n$ with $\sum_{n=0}^{\infty} |b_n| < \infty$.

16.6.c. Let $\varphi(\lambda) = \sum_{n=0}^{\infty} a_n \lambda^n$ be an entire function and let $x \in A \in \mathbf{BC}_e$. Prove that $\varphi(x) = \sum_{n=0}^{\infty} a_n x^n$, the symbol $\varphi(x)$ being defined by Theorem 16.4. It hence follows, in particular, that the notation of Theorem 16.4 coincides with that introduced in Definition 7.1.

§ 17. THE SPECTRUM AND THE JOINT SPECTRUM

In this section we deal with one of the most important concepts of the theory of Banach algebras, namely the concept of the spectrum and the joint spectrum of elements of an algebra. These notions have already been introduced in Section

10 (Definition 10.12). To recall the notion of the spectrum $\sigma(x)$ and some of its properties we formulate the following theorem.

17.1. THEOREM. *Let $A \in \mathbf{BC}_e$ and let $x \in A$. Then the following subsets of the complex plane C coincide*:

(i) $x^{\wedge}(\mathfrak{M})$,

(ii) $\{\lambda \in C: (x-\lambda e)^{-1} \notin A\}$.

Note that the set (ii) is defined also in the case of a non-commutative algebra, while the set (i) may be smaller or even empty. Therefore the spectrum $\sigma(x)$ is defined by (ii), provided A is any (possibly non-commutative) Banach algebra with unit. If A has no unit, the spectrum of an $x \in A$ is defined as the spectrum of this element in the algebra $A_1 = A \oplus \{\lambda e\}$. The spectrum of any element in an arbitrary Banach algebra always is a non-void compact set. Further note that the spectrum of x depends on the maximal commutative subalgebra A_0 of A containing x. In fact, if the elements $(x-\lambda e)$ are invertible in A, then their inverses commute with x as well as with any element commuting with x, so they are in A_0. However, if A_0 is a subalgebra of A and $x \in A_0$, the spectrum of x in A_0 need not coincide with the spectrum in A. Therefore we shall write specifically $\sigma_{A_0}(x)$ or $\sigma_A(x)$ whenever necessary.

We now prove some theorems concerning spectra, useful in further applications.

17.2. THEOREM. *Let A_0 be a subalgebra of an algebra $A \in \mathbf{BC}_e$. Then*

$$\sigma_A(x) \subset \sigma_{A_0}(x), \tag{17.2.1}$$

while the reverse inclusion holds for the boundaries of those spectra:

$$\partial\sigma_A(x) \supset \partial\sigma_{A_0}(x). \tag{17.2.2}$$

PROOF. Inclusion (17.2.1) follows from the fact that the restriction of any functional $f \in \mathfrak{M}(A)$ to A_0 is an element of $\mathfrak{M}(A_0)$. Now let $\lambda_0 \in \partial\sigma_{A_0}(x)$. As shown in the proof of Theorem 14.2, the element $x-\lambda_0 e$ is a topological divisor of zero in A_0, hence cannot be invertible in any superalgebra of A_0. Thus $\lambda_0 \in \sigma_A(x)$ and $\partial\sigma_{A_0}(x) \subset \sigma_A(x)$. Finally, since (17.2.1) can be written as $C \setminus \sigma_{A_0}(x) \subset C \setminus \sigma_A(x)$, then

$$\sigma_{A_0}(x) \cap \overline{C \setminus \sigma_{A_0}(x)} = \partial\sigma_{A_0}(x) \subset \sigma_A(x) \cap \overline{C \setminus \sigma_A(x)} = \partial\sigma_A(x)$$

and (17.2.2) follows.

17.3. REMARK. Theorem 17.2 is valid for arbitrary (possibly non-commutative or without units) Banach algebras since the spectra in question depend but on some commutative algebras with units (see the remarks preceding Theorem 17.2).

17.4. COROLLARY. *If $x \in A_0 \subset A$ and the spectrum $\sigma_{A_0}(x)$ has no interior points, then $\sigma_{A_0}(x) = \sigma_A(x)$.*

PROOF. In view of (17.2.1) and (17.2.2), we have

$$\sigma_{A_0}(x) \subset \partial\sigma_{A_0}(x) \subset \partial\sigma_A(x) \subset \sigma_A(x) \subset \sigma_{A_0}(x).$$

17.4.a. EXERCISE. Show by an example that it can happen that $\sigma_A(x) = \partial\sigma_A(x)$ and $\sigma_{A_0}(x) \neq \partial\sigma_{A_0}(x)$ for some $x \in A_0 \subset A$. Thus the assumption on the spectrum $\sigma_{A_0}(x)$ in Corollary 17.4 cannot be replaced by a similar one involving $\sigma_A(x)$.

17.5. COROLLARY. *If $x \in A_0 \subset A$ and $\sigma_A(x)$ does not disconnect the plane $\mathbf{C}$, then $\sigma_{A_0}(x) = \sigma_A(x)$.*

PROOF. Suppose that there exists a point $\lambda_0 \in \sigma_{A_0}(x) \setminus \sigma_A(x)$. Since $\sigma_A(x)$ does not disconnect the plane, then there exists an arc disjoint with $\sigma_A(x)$ and connecting λ_0 with the point at infinity. This arc must contain at least one boundary point of $\sigma_{A_0}(x)$, hence a point of $\sigma_A(x)$, by (17.2.2). This contradiction implies the inclusion $\sigma_{A_0}(x) \subset \sigma_A(x)$ which, in view of (17.2.1), gives the desired equality.

17.6. COROLLARY. *If $x \in A$ and if $\sigma_A(x)$ has no interior points and does not disconnect the plane $\mathbf{C}$ (in particular, if $\sigma_A(x)$ lies on the real axis), then $\sigma_A(x)$ is also the spectrum of x in any superalgebra of A and in any subalgebra of A containing x.*

REMARK. Corollaries 17.4–17.6 are valid (similarly as in Theorem 17.2 — see Remark 17.3) for arbitrary (possibly non-commutative or without units) Banach algebras (however, the unit, if it exists must be common for all sub- and superalgebras in question).

Let us prove a theorem on spectra, useful in the theory of *-algebras.

17.7. THEOREM. *If A is a non-commutative Banach algebra and if $x, y \in A$, then $\sigma(xy) \cup \{0\} = \sigma(yx) \cup \{0\}$.*

PROOF. In view of the symmetry, it suffices to show that if $\lambda \notin \sigma(xy) \cup \{0\}$, then $\lambda \notin \sigma(yx) \cup \{0\}$. Replacing x by $-\lambda^{-1}x$ we reduce the problem to a proof of the fact that if $-1 \notin \sigma(xy)$, then $-1 \notin \sigma(yx)$ or, which is the same, that the invertibility of $e+xy$ implies that of $e+yx$. So it suffices to verify the equality

$$(e+yx)^{-1} = e-y(e+xy)^{-1}x,$$

which is left to the reader as an easy exercise.

17.7.a. EXERCISE. Show that $\sigma(xy) \neq \sigma(yx)$ can happen. Employ the algebra $B(l_1)$ and the operators $x, y \in B(l_1)$ given by

$$x(\alpha_1, \alpha_2, \ldots) = (\alpha_2, \alpha_3, \ldots),$$

$$y(\alpha_1, \alpha_2, \ldots) = (0, \alpha_1, \alpha_2, \ldots).$$

We shall now be concerned with joint spectra. We recall the definition: the joint spectrum of a system of elements $x_1, \ldots, x_n$ of an algebra $A \in \mathbf{BC}_e$ is the compact set

$$\sigma_A(x_1, \ldots, x_n) = \{(x_1\hat{}(M), \ldots, x_n\hat{}(M)) \in \mathbf{C}^n : M \in \mathfrak{M}(A)\}.$$

17.8. THEOREM. *Let $A \in \mathbf{BC}_e$ and let $x_1, \ldots, x_n \in A$. Then*

$$\sigma_A(x_1, \ldots, x_n) = \{(\lambda_1, \ldots, \lambda_n) \in C^n: \text{for any } y_1, \ldots, y_n \in A \text{ the element } \sum_{i=1}^{n} y_i(x_i - \lambda_i e) \text{ is not invertible in } A\}. \tag{17.8.1}$$

PROOF. Indeed, the set of all elements y of the form $y = \sum_{i=1}^{n} y_i(x_i - \lambda_i e)$ with $y_i \in A$ is either a proper ideal or is the whole algebra A. Consequently, if this set consists entirely of singular elements, then it is an ideal contained in some maximal ideal $M_0 \in \mathfrak{M}(A)$. Thus

$$f_{M_0}(y) = \sum_{i=1}^{n} f_{M_0}(y_i)\,[f_{M_0}(x_i) - \lambda_i] = 0$$

for any choice of y_i's. Putting $y_i = \delta_{ij} e$, $j = 1, \ldots, n$, we get $f_{M_0}(x_j) = \lambda_j$, so that $(\lambda_1, \ldots, \lambda_n) = (f_{M_0}(x_1), \ldots, f_{M_0}(x_n)) \in \sigma(x_1, \ldots, x_n)$ which means that the set occurring on the left-hand side of (17.8.1) is contained in the set on the right-hand side. For the opposite inclusion, observe that if $(\lambda_1, \ldots, \lambda_n) \in \sigma(x_1, \ldots, x_n)$, then there exists a functional $f \in \mathfrak{M}(A)$ such that $f(x_i) = \lambda_i$, $i = 1, \ldots, n$. It follows that $f(y) = 0$ for all $y = \sum_{i=1}^{n} y_i(x_i - \lambda_i e)$ and so every such y is not invertible in A, which gives the desired inclusion.

17.8.a. EXERCISE. The right-hand side of (17.8.1) has a sense also in the non-commutative case. Letting the joint spectrum be defined by (17.8.1) find out which of the theorems of this section remain valid in the general case.

Note that every compact subset $K \subset C^n$ is the joint spectrum of a system of elements of some Banach algebra A. Namely, if $A = C(K)$ and $x_i = x_i(\lambda) \equiv \lambda_i$, $i = 1, \ldots, n$, where $\lambda = (\lambda_1, \ldots, \lambda_n)$ is the variable in C^n, then $\mathfrak{M}(A) = K$ and

$$\begin{aligned}\sigma(x_1, \ldots, x_n) &= \{(\hat{x_1}(\lambda), \ldots, \hat{x_n}(\lambda)) \in C^n: \lambda \in K\} \\ &= \{(\lambda_1, \ldots, \lambda_n) \in C^n: \lambda \in K\} = K.\end{aligned}$$

Observe that the algebra $A = C(K)$ is not generated by $x_1, \ldots, x_n$ whenever K has non-void interior. In fact, in this case the algebra generated by $x_1, \ldots, x_n$ consists of functions holomorphic in the interior of K. We shall discuss this case in detail.

17.9. DEFINITION. Let K be a non-void compact subset of C^n. The *polynomially convex hull of* K is the set

$$P(K) = \bigcap_{W} \{\lambda \in C^n: |W(\lambda)| \leqslant \max_{\mu \in K} |W(\mu)|\}, \tag{17.9.1}$$

the intersection being taken over all complex polynomials W.

Clearly $K \subset P(K)$ and $P(K)$ is compact. In view of the equality

$$\{\lambda \in C^n: |W(\lambda)| = 0\} = \bigcap_{k=1}^{\infty} \left\{\lambda \in C^n: \left|W(\lambda) + \frac{1}{k}\right| \leqslant \frac{1}{k}\right\}$$

we may restrict our attention to those polynomials whose maximum modulus on K is positive. Dividing those polynomials by the corresponding maximal values we get an alternative description of $P(K)$:

$$P(K) = \bigcap_{\substack{\max|W(\mu)|=1\\ \mu\in K}} \{\lambda \in C^n: |W(\lambda)| \leqslant 1\}. \tag{17.9.2}$$

A compact set $K \subset C^n$ is called *polynomially convex* if $K = P(K)$.

17.9.a. EXERCISE. Show that polynomially convex connected subsets of the complex plane C are simply connected.

17.10. THEOREM. *Let $A \in \mathbf{BC}_e$. If A is generated by elements $x_1, \ldots, x_n$, then the joint spectrum of those elements is polynomially convex.* Conversely, *every polynomially convex subset of C^n is the joint spectrum of a system of n generators of some Banach algebra.*

PROOF. If $A = [x_1, \ldots, x_n]$, then $\sigma(x_1, \ldots, x_n) = \mathfrak{M}(A)$ in view of 10.12. Let $K = \sigma(x_1, \ldots, x_n)$ and let $\lambda_0 \in P(K)$, $\lambda_0 = (\lambda_1, \ldots, \lambda_n) \in C^n$. For any polynomial W of n complex variables we have $|W(\lambda_0)| \leqslant \max_{\mu\in K}|W(\mu)|$. Write $y_W = W(x_1, \ldots, x_n) \in A$ and $f_0(y_W) = W(\lambda_0)$. The elements y_W are dense in A; the functional f_0 is multiplicative and linear on the set of y_W's. Moreover, f_0 is continuous since

$$|f_0(y_W)| = |W(\lambda_0)| \leqslant \max_{\mu\in K}|W(\mu)| = \max_{f\in\mathfrak{M}}|f(W(x_1, \ldots, x_n))|$$
$$= ||W(x_1, \ldots, x_n)||_s \leqslant ||W(x_1, \ldots, x_n)|| = ||y_W||$$

and consequently f_0 extends to an element of $\mathfrak{M}(A)$ whence $\lambda_0 \in \sigma(x_1, \ldots, x_n) = K$. Thus $P(K) \subset K$ and K is polynomially convex.

Conversely, if $K = P(K)$, then the desired algebra A may be defined as the completion in the norm $||W|| = \max_{\lambda\in K}|W(\lambda)|$ of the algebra of all polynomials of n variables. Indeed, every point $\lambda \in K$ defines a continuous multiplicative linear functional $f(W) = W(\lambda)$ on the algebra of polynomials W; this functional extends by continuity to an element of $\mathfrak{M}(A)$. On the other hand, if $f \in \mathfrak{M}(A)$, then, writing $\mu = (\mu_1, \ldots, \mu_n)$, where $\mu_i = f(W_i)$, $W_i(\lambda) \equiv \lambda_i$, $i = 1, \ldots, n$, we get $f(W) = W(\mu)$ for any polynomial W. The point μ certainly is in K since otherwise there exists a polynomial W with $|W(\mu)| > ||W||$. Replacing W by a suitable multiple of it, we may assume that $|W(\mu)| > 1 > ||W||$, whence $W^n \to 0$ while $f(W^n) \to \infty$, a contradiction to the continuity of f. We have thus proved that $K = \mathfrak{M}(A)$. The algebra A is generated by the n monomials W_i, $i = 1, \ldots, n$.

Now we shall discuss the continuity of the operation $x \to \sigma(x)$.

17.11. DEFINITION. Let M be a metric space with a metric $r(x, y)$ and let $\Phi\colon M \to 2^{C^n}$ be a function which maps M into the family of all subsets of the space C^n. The function Φ is said to be *continuous at a point* $x \in M$ if for every neighbourhood of zero V in C^n a number $\delta > 0$ exists such that the inequality $r(x, y) < \delta$ implies

$$\Phi(x) \subset \Phi(y) + V, \tag{17.11.1}$$

$$\Phi(y) \subset \Phi(x) + V, \tag{17.11.2}$$

where

$$\Phi(z) + V = \{p \in C^n : p = p_1 + p_2,\ p_1 \in \Phi(z), p_2 \in V\}.$$

If $r(x, y) < \delta$ implies (17.11.2) only, then Φ is called *upper-semi-continuous*.

17.12. THEOREM. *Let $A \in \mathbf{BC}_e$. Then the joint spectrum of a system of elements $x_1, \ldots, x_n \in A$, regarded as a mapping of the product $A \times \ldots \times A$ into the family of all subsets of C^n, is continuous.*

PROOF. We first prove (17.11.2). Suppose that (17.11.2) does not hold. Then there exist elements $x_1^{(0)}, \ldots, x_n^{(0)} \in A$ and an open neighbourhood of zero V in C^n such that for any positive integer k there exist elements $x_1^{(k)}, \ldots, x_n^{(k)} \in A$ and a functional $f_k \in \mathfrak{M}(A)$ such that

$$\|x_i^{(k)} - x_i^{(0)}\| < \frac{1}{k}, \quad i = 1, \ldots, n, \tag{17.12.1}$$

and

$$(f_k(x_1^{(k)}), \ldots, f_k(x_n^{(k)})) \notin \sigma(x_1^{(0)}, \ldots, x_n^{(0)}) + V.$$

Since $|f_k(x_i^{(k)})| \leqslant \|x_i^{(0)}\| + 1$ for $i = 1, \ldots, n$, then the sequences $\{f_k(x_i^{(k)})\}_{k=1}^{\infty}$ are bounded and may be assumed to converge (we pass to subsequences if necessary). Let $\lambda_i = \lim_k f_k(x_i^{(k)})$. Since V is open, so is $\sigma(x_1^{(0)}, \ldots, x_n^{(0)}) + V$. Consequently the vector $\lambda = (\lambda_1, \ldots, \lambda_n)$ does not belong to $\sigma(x_1^{(0)}, \ldots, x_n^{(0)}) + V$, neither, of course, to $\sigma(x_1^{(0)}, \ldots, x_n^{(0)})$ and, by Theorem 17.8, there exist elements $y_1, \ldots, y_n \in A$ such that

$$\sum_{i=1}^{n} y_i(x_i^{(0)} - \lambda_i e) = e. \tag{17.12.2}$$

Since $x_i^{(0)} = \lim_k x_i^{(k)}$ and the set of all invertible elements of A is open in A, then we can replace the elements $x_i^{(0)} - \lambda_i e$ in the sum (17.12.2) by elements $x_i^{(k)} - f_k(x_i^{(k)})e$ (with sufficiently large k), preserving invertibility of the sum

$$y = \sum_{i=1}^{n} y_i[x_i^{(k)} - f_k(x_i^{(k)})e].$$

However, this is impossible, since $f_k(y) = 0$.

Inclusion (17.11.1) will also be proved indirectly. Suppose that (17.11.1) does not hold. Then there exist elements $x_1^{(0)}, \ldots, x_n^{(0)} \in A$ and an open neighbourhood of zero $V = \{\lambda \in C^n\colon |\lambda| < \varepsilon\}$ such that for any positive integer k there exist elements $x_1^{(k)}, \ldots, x_n^{(k)} \in A$ satisfying (17.12.1) and a functional $f_k \in \mathfrak{M}$ such that

$$(f_k(x_1^{(0)}), \ldots, f_k(x_n^{(0)})) \notin \sigma(x_1^{(k)}, \ldots, x_n^{(k)}) + V.$$

We have $\lim\limits_k x_i^{(k)} = x_i^{(0)}$. Similarly as in the proof of (17.11.2) we may assume that the limits $\lambda_i^{(0)} = \lim\limits_k f_k(x_i^{(0)})$ exist for $i = 1, \ldots, n$. Writing $\lambda^{(k)} = (f_k(x_1^{(0)}), \ldots$ $\ldots, f_k(x_n^{(0)}))$ we have $\lambda^{(k)} - \lambda^{(0)} \in \frac{1}{2}V$ for large k, say, for $k > K$ and, since $\lambda^{(k)} \notin \sigma(x_1^{(k)}, \ldots, x_n^{(k)})$, so

$$\lambda^{(0)} \notin \sigma(x_1^{(k)}, \ldots, x_n^{(k)}) + \tfrac{1}{2}V \quad \text{for } k > K. \tag{17.12.3}$$

Since $\lambda^{(k)} \in \sigma(x_1^{(0)}, \ldots, x_n^{(0)})$ and $\lambda^{(0)} = \lim\limits_k \lambda^{(k)}$, then $\lambda^{(0)} \in \sigma(x_1^{(0)}, \ldots, x_n^{(0)})$ and there exists a functional $f_0 \in \mathfrak{M}$ such that $f_0(x_i^{(0)}) = \lambda_i^{(0)}$, $i = 1, \ldots, n$. On the other hand, relation (17.12.3) is equivalent to the inequality

$$\max_{f \in \mathfrak{M}} |\lambda^{(0)} - (f(x_1^{(k)}), \ldots, f(x_n^{(k)}))| \geqslant \tfrac{1}{2} \quad \text{for } k > K,$$

in particular

$$|(f_0(x_1^{(0)}), \ldots, f_0(x_n^{(0)})) - (f_0(x_1^{(k)}), \ldots, f_0(x_n^{(k)}))| \geqslant \tfrac{1}{2}\varepsilon \quad \text{for } k > K,$$

which is impossible, as $x_i^{(0)} = \lim\limits_k x_i^{(k)}$.

REMARK. The above theorem is not valid in non-commutative algebras even in the case of spectra of single elements (see the example below); but, in any case, the spectrum, regarded as an operation, always is upper-semi-continuous (Rickart [1], p. 35).

17.13. EXAMPLE (Rickart [1], p. 282). Let A be the Banach algebra of all endomorphisms of the space l^2. For an $x = (x_n)_{n=1}^{\infty} \in l^2$ let $Tx = (\alpha_n x_{n+1})$ with $\alpha_n = \exp(-m)$, where m is the exponent occurring in the (unique) decomposition $n = 2^m(2l+1)$. It is not difficult to see that $\|T\| = \sup\limits_k |\alpha_k|$ and $\|T^n\| = \sup\limits_k |\alpha_k \alpha_{k+1} \ldots \alpha_{k+n}|$ since $T^n x = (\alpha_k \alpha_{k+1} \ldots \alpha_{k+n} x_{k+n})_{k=1}^{\infty}$. By the definition of α_n we have

$$\alpha_1 \alpha_2 \ldots \alpha_{2^k-1} = \prod_{j=1}^{k-1} \exp(-j2^{k-j-1}),$$

thus

$$(\alpha_1 \alpha_2 \ldots \alpha_{2^k-1})^{1/2^k-1} > \Big(\prod_{j=1}^{k-1} \exp(-j/2^{j+1})\Big)^2$$

$$= \exp\Big(-2\sum_{j=1}^{k-1} j/2^{j+1}\Big).$$

Writing $\beta = \sum_{j=1}^{\infty} j/2^{j+1}$ we have $\lim \sqrt[n]{\|T^n\|} \geqslant \exp(-2\beta)$ so that $\|T\|_s > 0$ and the spectrum $\sigma(T)$ contains non-zero elements. Let $e_k = (\delta_{ik})_{i=1}^{\infty}$, $k = 1, 2, \ldots$, be the usual basis for l^2; we define operators T_n, $n = 1, 2, \ldots$, by

$$T_n e_k = \begin{cases} 0 & \text{for } k = 2^n(2l+1), \\ \alpha_k e_{k+1} & \text{for } k \neq 2^n(2l+1). \end{cases}$$

Since $T_n^{2^{n+1}} = 0$, then $\|T_n\|_s = 0$ and $\sigma(T_n) = \{0\}$. On the other hand, since

$$(T-T_n)e_k = \begin{cases} e^{-n}e_{k+1} & \text{for } k = 2^n(2l+1), \\ 0 & \text{for } k \neq 2^n(2l+1), \end{cases}$$

then $\|T-T_n\| = e^{-n}$ and $T = \lim T_n$. It follows that a sequence of elements with spectra reducing to $\{0\}$ may converge to an element with non-zero spectrum.

We conclude this section with some remarks concerning joint spectra of infinitely many elements.

17.14. Definition. Let $F \subset A \in \mathbf{BC}_e$. The *joint spectrum $\sigma(F)$ of the family of elements F* is the set

$$\sigma_A(F) = \{(x^\wedge(M))_{x\in F} \in \mathbf{C}^F \colon M \in \mathfrak{M}(A)\}.$$

It is a subset of the product of copies of the plane $\mathbf{C}$, indexed by the set F. Since $|x^\wedge(M)| \leqslant \|x\|$, then

$$\sigma_A(F) \subset \prod_{x\in F} K(0, \|x\|) \subset \mathbf{C}^F,$$

where $K(0, r) = \{\lambda \in \mathbf{C} \colon |\lambda| \leqslant r\}$. The spectrum $\sigma(F)$ is thus a compact set.

17.14.a. Exercise. Prove that if $A = [F]$, then the spectrum $\sigma_A(F) = \mathfrak{M}(A)$.

17.14.b. Exercise. Prove that if A_0 is a subalgebra of $A \in \mathbf{BC}_e$ and $F \subset A_0$, then $\sigma_A(F) \subset \sigma_{A_0}(F)$.

17.14.c. Exercise. Prove that if $A = [F]$, then the spectrum $\sigma_A(F)$ is a polynomially convex subset of the space $\mathbf{C}^F$, i.e.

$$\sigma_A(F) = \bigcap_{\nu} \{\xi \in \mathbf{C}^F \colon |p_\nu(\xi)| \leqslant 1\},$$

where p_ν are polynomials in a finite number of variables, which means that each $p_\nu(\xi)$ is a function depending (polynomially) on finitely many coordinates only, and $\sup_{\xi\in\sigma(F)} |p_\nu(\xi)| = 1$.

§ 18. A CHARACTERIZATION OF MAXIMAL IDEALS IN BANACH ALGEBRAS

In this section we make use of analytic functions in proving the following theorem together with some of its consequences.

18.1. Theorem. *Let $A \in \mathbf{BC}_e$. A linear subspace $X \subset A$ of codimension* 1 *is a maximal ideal in A if and only if it consists of non-invertible elements.*

Clearly any maximal ideal satisfies the above condition. We remark also that such a subspace X must necessarily be closed, since it is disjoint from an open set $G(A)$. We reformulate now the problem in the language of linear functionals on A. Since $\operatorname{codim} X = 1$ and $e \notin X$, there is a unique linear functional f on A such that

$$f(e) = 1 \tag{18.1.1}$$

and

$$f(x) = 0 \tag{18.1.2}$$

for all $x \in X$. Such a functional can be characterized by

$$f(x) \in \sigma(x) \tag{18.1.3}$$

for all $x \in A$. In fact, if $f \in A'$ and f satisfies (18.1.3), then clearly (18.1.1) holds, and, by Theorem 17.1 the subspace $X = \{x: f(x) = 0\} \subset A$ of codimension 1 consists of non-invertible elements. On the other hand, if f satisfies (18.1.1) and (18.1.2) and X consists of non-invertible elements, $\operatorname{codim} X = 1$, then for any $x \in A$ the element $y = x - f(x)e$ is in X, so it is non-invertible and $0 \in \sigma(x - f(x)e) = \sigma(x) - f(x)$, which implies (18.1.3).

Thus Theorem 18.1 is equivalent with the following

18.2. THEOREM. *Let $A \in \mathbf{BC}_e$. Then a functional $f \in A'$ belongs to $\mathfrak{M}(A)$ if and only if* (18.1.3) *holds for all $x \in A$.*

PROOF. If $f \in \mathfrak{M}(A)$, then by the definition of the spectrum (18.1.3) holds true. Suppose now that (18.1.3) is satisfied for all $x \in A$, hence also (18.1.1). Fix an element $x \in A$ and consider elements $\exp \lambda x$, where λ is a complex variable (cf. Exercise 16.5.c). We put

$$\varphi(\lambda) = f(\exp \lambda x). \tag{18.2.1}$$

It can be easily verified that φ is an entire function. Since, by (18.1.3) $\varphi(\lambda) \neq 0$ for all $\lambda \in \boldsymbol{C}$, it may be written in the form

$$\varphi(\lambda) = \exp[\psi(\lambda)],$$

where $\psi(\lambda)$ is an entire function. We have also

$$|\varphi(\lambda)| \leqslant \|f\| \left\| \sum_{0}^{\infty} \frac{\lambda^n x^n}{n!} \right\| \leqslant \|f\| \sum_{0}^{\infty} \frac{|\lambda|^n \|x\|^n}{n!} = \|f\| \exp(|\lambda|\, \|x\|)$$

and so, by a classical theorem of Hadamard (cf. e.g. Titchmarsh [1]) $\psi(\lambda)$ must be a polynomial of first degree, $\psi(\lambda) = \alpha + \beta\lambda$. Thus $\varphi(\lambda) = \exp\alpha \exp\beta\lambda$, and setting $\lambda = 0$ we see by (18.1.1) that $\exp\alpha = 1$, so $\varphi(\lambda) = \exp\beta\lambda$. We then have the expansion

$$\varphi(\lambda) = \sum_{0}^{\infty} \frac{\beta^n}{n!} \lambda^n. \tag{18.2.2}$$

On the other hand, the definition of $\varphi(\lambda)$ implies the expansion

$$\varphi(\lambda) = f\left(\sum_{0}^{\infty} \frac{x^n \lambda^n}{n!}\right) = \sum_{0}^{\infty} \frac{f(x^n)}{n!} \lambda^n. \tag{18.2.3}$$

Comparing the coefficients in expansions (18.2.2) and (18.2.3) we have $f(x^n) = \beta^n$ and in particular

$$f(x^2) = \beta^2 = f(x)^2$$

for any $x \in A$. This implies

$$f(xy) = f(\tfrac{1}{2}[(x+y)^2 - x^2 - y^2]) = \tfrac{1}{2}[f(x+y)^2 - f(x)^2 - f(y)^2] = f(x)f(y)$$

and so $f \in \mathfrak{M}(A)$.

18.2.a. EXERCISE. Prove that the function $\varphi(\lambda)$ given by formula (18.2.1) is an entire function.

18.3. REMARK. Theorem 18.2 is false for real Banach algebras, which can easily be seen by setting $A = C(0, 1)$ and $f(x) = \int_0^1 x(t)\,dt$.

As a corollary we obtain a theorem on multiplicativity of measures possessing mean-value property.

18.4. THEOREM. *Let Ω be a compact Hausdorff space and let A be a complex algebra of complex-valued continuous functions defined on Ω, containing the constants. Let μ be a Radon measure on Ω, such that for any function $x \in A$*

$$\int x\,d\mu = x(p_x), \tag{18.4.1}$$

where p_x is a point of Ω depending on x (the mean-value property). Then the measure μ is multiplicative with respect to A, i.e.

$$\int xy\,d\mu = \int x\,d\mu \int y\,d\mu \tag{18.4.2}$$

for all $x, y \in A$.

PROOF. Let $\bar{A}$ be the completion of A in the sup norm on Ω, so $\bar{A} \in \mathbf{BC}_e$. By formula (18.4.1) the functional

$$f(x) = \int x\,d\mu$$

is continuous with respect to the sup norm, so it can be extended to $\bar{A}$ by continuity. If $x \in A$, then $x = \lim x_n$, $x_n \in A$, and $f(x) = \lim f(x_n) = \lim x_n(p_n)$, where p_n are suitable points in Ω.

So

$$|x(p_n) - f(x)| \leqslant |x(p_n) - x_n(p_n)| + |f(x_n) - f(x)| \to 0$$

which means, by the compactness of Ω, that there is a point $p \in \Omega$ such that $x(p) = f(x)$. Formula (18.2.3) holds then for all $x \in \bar{A}$ and the conclusion is now a consequence of Theorem 18.2.

18.4.a. EXERCISE. Prove Theorem 18.4 without the assumption that A contains the constants.

We shall now prove a generalization of a part of Theorem 18.2 replacing there the complex plane by an arbitrary commutative semi-simple Banach algebra with unit.

18.5. THEOREM. *Let $A_1, A_2 \in \mathbf{BC}_e$ and suppose that A_2 is a semi-simple algebra. If T is a linear map of A_1 into A_2 such that*

$$\sigma(Tx) \subset \sigma(x) \tag{18.5.1}$$

for any $x \in A_1$, then T is a multiplicative map, i.e.

$$Txy = Tx\,Ty \tag{18.5.2}$$

for all $x, y \in A_1$.

PROOF. Let $f \in \mathfrak{M}(A_2)$ and put

$$F(x) = f(Tx)$$

for $x \in A_1$, so F is a linear functional on A_1. We have also

$$F(x) = f(Tx) \in \sigma(Tx) \subset \sigma(x)$$

and so, by Theorem 18.2, $F \in \mathfrak{M}(A_1)$. It follows that

$$f(Txy) = F(xy) = F(x)F(y) = f(Tx)f(Ty) = f(TxTy).$$

Since this holds for any $f \in \mathfrak{M}(A_2)$ and A_2 is a semi-simple algebra, then $Txy = TxTy$ for all $x, y \in A_1$.

18.5.a. EXERCISE. Prove that Theorem 18.5 is false without the assumption of semi-simplicity of A_2.

The converse of Theorem 18.5 is true under an additional assumption that T sends the unit of A_1 into the unit of A_2. We may omit, however, the assumption that A_2 is a semi-simple algebra.

18.6. THEOREM. *Let $A_1, A_2 \in \mathbf{BC}_e$, and denote the units of A_1 and A_2 respectively by e_1 and e_2. Let T be a multiplicative linear map of A_1 into A_2, i.e. a linear map satisfying* (18.5.2). *Then for any $x \in A_1$ relation* (18.5.1) *holds, provided $Te_1 = e_2$.*

PROOF. We have

$$e_2 = Te_1 = Txx^{-1} = Tx\,Tx^{-1}$$

for any $x \in G(A_1)$. So if $x \in G(A_1)$, then $Tx \in G(A_2)$ and

$$(Tx)^{-1} = Tx^{-1}. \tag{18.6.1}$$

If $\lambda \notin \sigma(x)$, then $x - \lambda e_1 \in G(A_1)$, so $T(x-\lambda e_1) = Tx - \lambda e_2 \in G(A_2)$ and $\lambda \notin \sigma(Tx)$, which implies formula (18.5.1).

18.6.a. EXERCISE. Show by an example that a multiplicative Banach algebra need not satisfy (18.5.1) if it does not satisfy the relation $Te_1 = e_2$.

18.6.b. EXERCISE. Let $A_1, A_2 \in \mathbf{BC}_e$ and assume that they are semi-simple algebras. Let T be a linear map of A_1 onto A_2. Prove that T is anisomorphism between A_1 and A_2 if and only if $\sigma(x) = \sigma(Tx)$ for all $x \in A_1$.

We now consider the non-commutative case. First we need the following purely algebraic lemma.

18.7. LEMMA. *Let A be a real or complex algebra with unit e. Let f be a linear functional on A satisfying*

(18.7.1) $$f(e) = 1$$

and

(18.7.2) $$f(x^2) = f(x)^2$$

for all $x \in A$. Then f is a multiplicative functional on A, i.e. $f(xy) = f(x)f(y)$ for all $x, y \in A$.

PROOF. By (18.7.2) we have

$$f[(x+y)^2] = [f(x)+f(y)]^2$$

or

(18.7.3) $$f(xy+yx) = 2f(x)f(y)$$

for all $x, y \in A$. It follows that if we put

$$x\Box y = \tfrac{1}{2}(xy+yx),$$

we obtain a (non-associative) multiplication on A such that

$$f(x\Box y) = f(x)f(y)$$

for all $x, y \in A$. Consequently

$$f[(x\Box y)\Box z] = f(x)f(y)f(z),$$

that is equivalent to

(18.7.4) $$f(xyz+zxy+yxz+zyx) = 4f(x)f(y)f(z),$$

$x, y, z \in A$. Also

$$f[y\Box(z\Box x)-(y\Box z)\Box x] = 0,$$

which is equivalent with

(18.7.5) $$f(xyz+zyx) = f(zxy+yxz)$$

for any $x, y, z \in A$. From (18.7.4) and (18.7.5) it follows that

$$f(xyz+zyx) = 2f(x)f(y)f(z).$$

Substituting here $x = z$ we obtain

(18.7.6) $$f(xyx) = f(x)^2 f(y)$$

for $x, y \in A$. We want to show that

(18.7.7) $$f(xy) = f(yx)$$

for all $x, y \in A$. Suppose, on the contrary, that for some $x_0, y_0 \in A$ we have

$$f(x_0y_0-y_0x_0) = C \neq 0.$$

Taking here y_0/C instead of y_0 we may assume

(18.7.8) $$f(x_0 y_0 - y_0 x_0) = 1.$$

It is also clear that the above relation also holds if we take instead of x_0 any element of the form $x_0 + \alpha e$, where α is a scalar. We may therefore assume also that

(18.7.9) $$f(x_0) = 0$$

taking instead of x_0 the element $x_0 - f(x_0)e$.

By (18.7.3) and (18.7.9)

$$f(x_0 y_0) + f(y_0 x_0) = 0$$

and so by (18.7.8)

(18.7.10) $$f(x_0 y_0) = \tfrac{1}{2}, \quad f(y_0 x_0) = -\tfrac{1}{2}.$$

On the other hand, by (18.7.2), (18.7.6), (18.7.8), (18.7.9) and (18.7.10) we have

$$\begin{aligned} 1 &= f[(x_0 y_0 - y_0 x_0)^2] \\ &= f[(x_0 y_0)^2 + (y_0 x_0)^2 - x_0 y_0^2 x_0 - y_0 x_0^2 y_0] \\ &= \tfrac{1}{4} + \tfrac{1}{4} - 2f(x_0)^2 f(y_0)^2 = \tfrac{1}{2}, \end{aligned}$$

which is a contradiction proving formula (18.7.7). By (18.7.3) and (18.7.7) we obtain the desired result $f(xy) = f(x)f(y)$.

18.7.a. Exercise. Check all calculations ommited in the proof of Lemma 18.7.

18.8. Corollary. *If A is a real or complex algebra without unit, and if f is a linear functional on A satisfying* (18.7.2), *then f is a multiplicative linear functional on A.*

This proof is obtained by considering the algebra A_1, $A \oplus \{\lambda e\}$ and extending f onto A_1 by setting $f(e) = 1$.

18.8.a. Exercise. Give the proof of Corollary 18.8.

We now can have a stronger version of Theorem 18.2.

18.9. Theorem. *Let A be a complex Banach algebra. Then a functional $f \in A$ is a multiplicative linear functional on A if (and only if)*

(18.9.1) $$f(x) \in \sigma(x)$$

for all $x \in A$ (cf. remarks after Theorem 17.1).

Proof. We may assume that A possesses a unit e, otherwise we could consider the algebra $A_1 = A \oplus \{\lambda e\}$ and the extension of f onto this algebra, given by $f(e) = 1$, which clearly satisfies relation (18.9.1). Since for any commutative subalgebra $A_0 \subset A$ we have, by formula (17.2.1) (cf. Remark 17.3) $\sigma_A(x) \subset \sigma_{A_0}(x)$ for any $x \in A_0$, it follows by Theorem 18.2 that the restriction of f to an commutative subalgebra of A is a multiplicative linear functional. In particular $f(x^2) = f(x)^2$ for any $x \in A$ and so f is a multiplicative functional on A by Lemma 18.7.

18.9.a. EXERCISE. Let A be a Banach algebra without unit and let A_1, $A \oplus \{\lambda e\}$. Suppose that $f \in A'$ satisfies relation (18.9.1) on A. Prove that the extension of f onto A_1, given by $f(e) = 1$, also satisfies relation (18.9.1) on A_1.

18.10. COROLLARY. *Let $A \in \mathbf{B}_e$. A subspace $X \subset A$,* codim$X = 1$, *is a maximal two-sided ideal in A if and only if it consists of non-invertible elements.*

We see, by this corollary, that the absence of multiplicative linear functionals in many non-commutative Banach algebras with units is caused by the fact that the group $G(A)$ of invertible elements is so big that it intersects every linear subspace of codimension 1.

The results of this chapter were originally proved in Gleason [1], Kahane and Żelazko [1], and Żelazko [11]. For an another proof cf. also Browder [1].

§ 19. ANALYTIC FUNCTIONS OF SEVERAL COMPLEX VARIABLES

The results of Section 16 are generalized in this section to the case of functions of several variables. Thus we start with a number of concepts and theorems concerning such functions. The proofs may be found e.g. in the book Gunning and Rossi [1]. Throughout this section points of the space $\mathbf{C}^n$ will be denoted by single letters $z, w, \ldots$, where $z = (z_1, \ldots, z_n)$.

19.1. DEFINITION. An *open polydisc centered at $w \in \mathbf{C}^n$ with polyradius $(r_1, \ldots, r_n)$*, $r_i > 0$, is the set

$$\Delta(w; r_1, \ldots, r_n) = \{z \in \mathbf{C}^n \colon |z_i - w_i| < r_i\}.$$

A *closed polydisc* is the closure of an open polydisc and is defined by an analogous formula:

$$\bar{\Delta}(w; r_1, \ldots, r_n) = \{z \in \mathbf{C}^n \colon |z_i - w_i| \leqslant r_i\}.$$

19.2. DEFINITION. Let D be an open set in $\mathbf{C}^n$. A complex-valued function f defined on D is called *analytic* if for every $w \in D$ a polydisc $\Delta = \Delta(w; r_1, \ldots, r_n)$ exists such that

$$f(z) = \sum_{\nu_1, \ldots, \nu_n = 0}^{\infty} a_{\nu_1, \ldots, \nu_n} (z_1 - w_1)^{\nu_1} \ldots (z_n - w_n)^{\nu_n}, \tag{19.2.1}$$

the series being convergent for all $z \in \Delta$. It can be proved that a function is analytic if and only if it is analytic with respect to each variable z_i separately. If f is an analytic function in D and $\bar{\Delta}(w; r_1, \ldots, r_n)$ is a closed polydisc contained in D, then by a repeated application of the Cauchy integral formula we obtain the following integral representation:

$$(19.2.2) \qquad f(z) = \frac{1}{(2\pi i)^n} \int_{|w_1-\zeta_1|=r_1} \cdots \int_{|w_1-\zeta_n|=r_n} \frac{f(\zeta)d\zeta_1 \ldots d\zeta_n}{(\zeta_1-z_1)\ldots(\zeta_n-z_n)}$$

for any $z \in \Delta(w; r_1, \ldots, r_n)$.

Now suppose that $p_1(z), \ldots, p_k(z)$ are polynomials in n variables $z = (z_1, \ldots \ldots, z_n)$ and consider the open set

$$(19.3.1) \qquad D(p_1, \ldots, p_k; r_1, \ldots, r_{n+k}) = \{z \in C^n \colon |z_i| < r_i, |p_j(z)| < r_{j+n}, i = 1, \ldots, n, j = 1, \ldots, k\},$$

with $r_i > 0$, $i = 1, 2, \ldots, n+k$.

The following lemma is essential for the remainder of this section.

19.3. LEMMA. *For every function $f(z_1, \ldots, z_n)$ analytic in a set* (19.3.1) *and for every ε with $0 < \varepsilon < \min(r_1, \ldots, r_{n+k})$ there exists a function F analytic in the polydisc $\Delta(0; r_1-\varepsilon, \ldots, r_{n+k}-\varepsilon) \subset C^{n+k}$ such that*

$$(19.3.2) \qquad f(z_1, \ldots, z_n) = F(z_1, \ldots, z_n, p_1(z_1, \ldots, z_n), \ldots, p_k(z_1, \ldots, z_n)).$$

We refer to Gunning and Rossi [1] (Lemma F 7, p. 41) for a proof of this lemma in the case of equal r_i's. Lemma 18.3 is then obtained by an easy substitution.

We are now in a position to prove a generalization of Theorem 16.2 to analytic functions of several variables in the case when the considered elements $x_1, \ldots, x_n \in A$ are generators of the algebra in question.

19.4. LEMMA. *Let $A \in \mathbf{BC}_e$ and let $x_1, \ldots, x_n$ be a system of generators of the algebra A. Suppose that $f(z_1, \ldots, z_n)$ is a function analytic in an open set $U \subset C^n$ containing the joint spectrum $\sigma(x_1, \ldots, x_n)$. Then there exists an element $y \in A$ such that*

$$(19.4.1) \qquad y\hat{}(M) = f(\hat{x_1}(M), \ldots, \hat{x_n}(M))$$

for all $M \in \mathfrak{M}(A)$.

PROOF. The spectrum $\sigma(x_1, \ldots, x_n)$ is compact, the set U may thus be assumed to be bounded, so that

$$(19.4.2) \qquad \sigma(x_1, \ldots, x_n) \subset U \subset \Delta = \Delta(0; r_1, \ldots, r_n)$$

for some positive $r_1, \ldots, r_n$.

In virtue of Theorem 17.10 the spectrum $\sigma(x_1, \ldots, x_n)$ is polynomially convex, so we have

$$(19.4.3) \qquad \sigma(x_1, \ldots, x_n) = \bigcap_\nu F_\nu = \bigcap_\nu \Delta \cap F_\nu,$$

where $F_\nu = \{z \in C^n: |p_\nu(z)| \leqslant 1\}$ and p_ν are polynomials. Since

$$\Big(\bigcap_\nu \bar{\Delta} \cap F_\nu\Big) \cap (C^n \setminus U) = \emptyset$$

and the sets $\bar{\Delta} \cap F_\nu$ are compact, so there are finitely many indices $\nu_1, \ldots, \nu_k$ such that

$$\Big(\bigcap_{j=1}^{k} \bar{\Delta} \cap F_{\nu_j}\Big) \cap (C^n \setminus U) = \emptyset,$$

or

$$\bigcap_{j=1}^{k} \bar{\Delta} \cap F_{\nu_j} \subset U.$$

U is open, so there exist numbers $r_{n+1}, \ldots, r_{n+k} > 1$ such that, writing

$$G_{\nu_j} = \{z \in C^n: |p_{\nu_j}(z)| < r_{j+n}\},$$

we have the inclusion

(19.4.4) $$\bigcap_{j=1}^{k} G_{\nu_j} \cap \Delta \subset U.$$

We shall write for brevity $p_j(z)$ instead of $p_{\nu_j}(z)$, $j = 1, \ldots, k$. We have, by (19.4.2), (19.4.3), (19.4.4), (19.3.1),

(19.4.5) $$\sigma(x_1, \ldots, x_n) \subset D(p_1, \ldots, p_k; r_1, \ldots, r_{n+k}) \subset U.$$

Let $x_{n+i} = p_i(x_1, \ldots, x_n)$, $i = 1, \ldots, k$. It follows from (19.4.5) that

$$\sigma(x_1, \ldots, x_{n+k}) \subset \Delta(0; r_1, \ldots, r_{n+k}).$$

In virtue of Lemma 19.3 there exists a function $F(z_1, \ldots, z_{n+k})$ analytic in $\Delta(0; r_1-\varepsilon, \ldots, r_{n+k}-\varepsilon)$ satisfying (19.3.2), where ε is chosen so that

$$\sigma(x_1, \ldots, x_{n+k}) \subset \Delta(0; r_1-\varepsilon, \ldots, r_{n+k}-\varepsilon) \subset \bar{\Delta}(0; r_1-\varepsilon, \ldots, r_{n+k}-\varepsilon) \subset \Delta(0; r_1, \ldots, r_{n+k}).$$

Now, put

(19.4.6) $$y = \frac{1}{(2\pi i)^n} \int\limits_{|\zeta_1| = r_1-\varepsilon} \cdots \int\limits_{|\zeta_{n+k}| = r_{n+k}-\varepsilon} \frac{F(\zeta)\, d\zeta_1 \ldots d\zeta_{n+k}}{(\zeta_1 e - x_1) \ldots (\zeta_{n+k} e - x_{n+k})}.$$

The integrand is an A-valued function defined and continuous in the domain of integration, so integral (19.4.6) exists.

Let $M \in \mathfrak{M}(A)$. Applying the corresponding functional to (19.4.6) we obtain, by (19.3.2) and (19.2.2),

$$\begin{aligned} y^\wedge(M) &= F(\hat{x_1}(M), \ldots, \hat{x}_{n+k}(M)) \\ &= F[\hat{x_1}(M), \ldots, \hat{x_n}(M), p_1(\hat{x_1}(M), \ldots, \hat{x_n}(M)), \ldots, p_k(\hat{x_1}(M), \ldots, \hat{x_n}(M))] \\ &= f(\hat{x_1}(M), \ldots, \hat{x_n}(M)). \end{aligned}$$

An examination of the above proof shows that the assumption that A is generated by $x_1, \ldots, x_n$ is not essential here; the only important point was the existence of a set D defined by (19.3.1) and satisfying (19.4.5). We should be done if we could show that for any system of elements $x_1, \ldots, x_n \in A$ and for any open set U containing $\sigma(x_1, \ldots, x_n)$ there exists an open set (19.3.1) so that (19.4.5) holds. However, this is not the case and we shall apply here a trick due to Arens and Calderón [1]. Namely, we show how the problem can be reduced to considering finitely generated algebras only.

19.5. LEMMA. *Let $x_1, \ldots, x_n \in A \in \mathbf{BC}_e$ and $\sigma_A(x_1, \ldots, x_n) \subset U$, where U is an open set in $\mathbf{C}^n$. Then there exist elements $y_1, \ldots, y_p \in A$ such that*

$$\sigma_{[x_1, \ldots, x_n, y_1, \ldots, y_p]}(x_1, \ldots, x_n) \subset U.$$

PROOF. Choose an arbitrary point $z^{(0)} = (z_1^{(0)}, \ldots, z_n^{(0)}) \notin \sigma_A(x_1, \ldots, x_n)$. In view of Theorem 17.8 there exist elements $y_1, \ldots, y_n$ in A such that $\sum_{i=1}^{n} (x_i - z_i^{(0)}e)y_i = e$. The set of all invertible elements of the algebra $[x_1, \ldots, x_n, y_1, \ldots, y_n]$ is open, so there exists a neighbourhood $U_{z^{(0)}}$ of $z^{(0)}$ in $\mathbf{C}^n$ such that the element $\sum_{i=1}^{n} (x_i - z_i e)y_i$ is invertible for any $z = (z_1, \ldots, z_n) \in U_{z^{(0)}}$. Thus, by Theorem 17.8,

$$U_{z^{(0)}} \cap \sigma_{[x_1, \ldots, x_n, y_1, \ldots, y_n]}(x_1, \ldots, x_n) = \emptyset.$$

Hence also

$$U_{z^{(0)}} \cap \sigma_{A_0}(x_1, \ldots, x_n) = \emptyset$$

for every subalgebra A_0 such that $[x_1, \ldots, x_n, y_1, \ldots, y_n] \subset A_0 \subset A$ (cf. Exercise 17.14.b). Now let $Z = \{z \in \mathbf{C}^n : |z_i| \leqslant \|x_i\|\}$. Certainly $\sigma_{A_0}(x_1, \ldots, x_n) \subset Z$ for any such subalgebra A_0. $Z \setminus U$ is a compact set and $Z \setminus U \subset \bigcup_{z \notin \sigma(x_1, \ldots, x_n)} U_z$, so there exist finitely many points $z^{(1)}, z^{(2)}, \ldots, z^{(k)} \in \mathbf{C}^n$ such that $Z \setminus U \subset \bigcup_{i=1}^{k} U_{z^{(i)}}$. To each of the neighbourhoods $U_{z^{(i)}}$ there correspond elements $y_1^{(i)}, \ldots, y_{n_i}^{(i)}$ such that

$$U_{z^{(i)}} \cap \sigma_{[x_1, \ldots, x_n, y_1^{(i)}, \ldots, y_{n_i}^{(i)}]}(x_1, \ldots, x_n) = \emptyset.$$

Put

$$A_0 = [x_1, \ldots, x_n, y_1^{(1)}, \ldots, y_{n_1}^{(1)}, y_1^{(2)}, \ldots, y_{n_2}^{(2)}, \ldots, y_1^{(k)}, \ldots, y_{n_k}^{(k)}].$$

We have

$$\sigma_{A_0}(x_1, \ldots, x_n) \cap U_{z^{(i)}} = \emptyset, \quad i = 1, 2, \ldots, k,$$

whence

$$(Z \setminus U) \cap \sigma_{A_0}(x_1, \ldots, x_n) = \emptyset.$$

Finally, since $\sigma_{A_0}(x_1, \ldots, x_n) \subset Z$, so $\sigma_{A_0}(x_1, \ldots, x_n) \subset U$.

Now we can pass to the proof of the announced generalization of Theorem 16.2.

19.6. THEOREM. *Let $A \in \mathbf{BC}_e$ and let $x_1, \ldots, x_n \in A$. Then for every function f analytic in an open set U containing the joint spectrum $\sigma(x_1, \ldots, x_n)$ there exists an element $y \in A$ satisfying condition* (19.4.1).

PROOF. Let $A_0 = [x_1, \ldots, x_n, y_1, \ldots, y_k] \subset A$ be an algebra for which the inclusion $\sigma_{A_0}(x_1, \ldots, x_n) \subset U$ holds; such an algebra exists, as asserted by the preceding lemma. Let $\tilde{f}(z_1, \ldots, z_{n+k}) = f(z_1, \ldots, z_n)$. The function $\tilde{f}$ is analytic in the set $V = U \times \mathbf{C}^k \subset \mathbf{C}^{n+k}$. Of course, $\sigma_{A_0}(x_1, \ldots, x_n, y_1, \ldots, y_k) \subset V$, thus, by Lemma 19.4, there exists an element $y \in A$ such that

$$y\hat{}(M) = \tilde{f}(\hat{x_1}(M), \ldots, \hat{x_n}(M), \hat{y_1}(M), \ldots, \hat{y_k}(M))$$
$$= f(\hat{x_1}(M), \ldots, \hat{x_n}(M)).$$

The above theorem is usually referred to as the *Shilov theorem* (since its proof for finitely generated algebras is due to Shilov), or the *Shilov–Arens–Calderón theorem.*

It should be stressed that the function f is assumed to be globally analytic on $\sigma(x_1, \ldots, x_n)$, i.e. on a certain neighbourhood of this spectrum. It is not sufficient to require that f should be locally analytic, i.e. that for every point $z \in \sigma(x_1, \ldots$ $\ldots, x_n)$ there exist a neighbourhood U_z of z in $\mathbf{C}^n$ and a function f_z analytic in U_z and equal to f on $U_z \cap \sigma(x_1, \ldots, x_n)$.

We now give an outline of a proof of a theorem generalizing Theorem 16.4. Let K be a compact set in $\mathbf{C}^n$. The symbol $\mathrm{Hol}\,K$ denotes the algebra of all functions globally analytic on K. Convergent sequences in $\mathrm{Hol}\,K$ are defined as those sequences which consist of functions analytic in some common neighbourhood $U \supset K$ (depending on the sequence) and converge uniformly on U. It can be proved that if K is a polynomially convex set then the polynomials are dense in $\mathrm{Hol}\,K$ (cf. Gunning and Rossi [1], Theorem F 8).

19.7. THEOREM. *Let $A \in \mathbf{BC}_e$. There exists a unique assigning to every n-tuple $\mathbf{x} = (x_1, \ldots, x_n) \subset A$, and every function $f \in \mathrm{Hol}[\sigma(x_1, \ldots, x_n)]$ an element $f(x_1, \ldots, x_n) \in A$ in such a way that:*

(i) *for a fixed n-tuple $\mathbf{x}$ the map $\Phi_{\mathbf{x}}: f \to f(x_1, \ldots, x_n)$ is a continuous homomorphism of* $\mathrm{Hol}[\sigma(x_1, \ldots, x_n)]$ *into A, such that $\Phi_{\mathbf{x}}(1) = e$.*

(ii) *if $f \in \mathrm{Hol}[\sigma(x_1, \ldots, x_m)]$, $n > m$, and F is an extension of f to a member of* $\mathrm{Hol}[\sigma(x_1, \ldots, x_n)]$, *given by $F(z_1, \ldots, z_n) = f(z_1, \ldots, z_m)$, $z_i \in \mathbf{C}$, then $F(x_1, \ldots, x_m) = f(x_1, \ldots, x_m)$.*

SKETCH OF A PROOF. First suppose that $A = [x_1, \ldots, x_n]$ and define $\Phi_{\mathbf{x}}(f)$ as y, the element of A given by formula (19.4.6). It is not difficult to see that the element y does not depend on the choice of the polydisc Δ and that the desired assignment $1 \to e$, $z_1^{k_1} \ldots z_n^{k_n} \to x_1^{k_1} \ldots x_n^{k_n}$ is in fact obtained. Thus the polynomials $p(z_1, \ldots, z_n)$ are carried into the elements $p(x_1, \ldots, x_n)$; moreover, the mapping

$f \to \Phi_x(f)$ is continuous with respect to the topology of $\mathrm{Hol}(\sigma(x_1, \ldots, x_n))$. Since the polynomials are dense in this space, then Φ_x extends by continuity to the asserted homomorphism; this homomorphism is given by (19.4.6).

The assertion in the general case (when A is not generated by $(x_1, \ldots, x_n)$) as well as the uniqueness of Φ_x is proved similarly as was done in Theorems 19.6 and 16.4.

As an application of the above techniques we give here two theorems due to Rossi (another application will be given in the following section). But first we have to define some further function-theoretic notions.

19.8. Definition. Let D be an open subset of C^n and let (U_α) be an open covering of D. A *set of Cousin data* for the covering (U_α) is any family of functions $\{h_{\alpha,\beta}\}$ analytic in the sets $U_\alpha \cap U_\beta$ and satisfying the relation

(19.8.1) $$h_{\alpha\beta}+h_{\beta\gamma}+h_{\gamma\alpha} = 0$$

in $U_\alpha \cap U_\beta \cap U_\gamma$, provided this intersection is not empty.

In particular, $h_{\alpha\alpha} = 0$ and $h_{\alpha\beta} = -h_{\beta\alpha}$.

The *solution of the additive Cousin problem for the data* $\{h_{\alpha,\beta}\}$ is any family of functions $\{h_\alpha\}$ analytic in the sets U_α and satisfying the relation

(19.8.2) $$h_\alpha - h_\beta = h_{\alpha\beta}$$

in the intersection $U_\alpha \cap U_\beta$. It can be proved (Oka's theorem; cf. Gunning and Rossi [1], Theorem F 6) that if D is a set of the form (19.3.1), then for any open covering of D and for any Cousin data the additive Cousin problem is solvable.

19.9. Definition. Let $A \in \mathbf{BC}_e$. A closed subset $E \subset \mathfrak{M}(A)$ is called a *peak set* if there exists an element $x \in A$ such that $x\hat{\ }(M) = 1$ for $M \in E$ and $|x\hat{\ }(M)| < 1$ for $M \notin E$. Clearly, every peak set intersects the Shilov boundary $\Gamma(A)$.

A closed subset $E \subset \mathfrak{M}(A)$ is called a *local peak set* if there exist an open set U and an element $x_U \in A$ such that $E \subset U \subset \mathfrak{M}(A)$ and $\hat{x_U}(M) = 1$ for $M \in E$, $|\hat{x_U}(M)| < 1$ for $M \in \overline{U} \setminus E$.

The following theorem is called the *Rossi theorem on local peak sets*.

19.10. Theorem. *Let $A \in \mathbf{BC}_e$. Then every local peak set $E \subset \mathfrak{M}(A)$ is a peak set.*

Proof. Let E be a local peak set; let U be a neighbourhood of E and let $x \in A$ be an element with $x\hat{\ }(M) = 1$ for $M \in E$, $|x\hat{\ }(M)| < 1$ for $M \in \overline{U} \setminus E$, according to the definition; we fix such an x. Since E is compact, so we may assume that U is a finite union of neighbourhoods (10.1.1) with ε replaced by 1 (this always may be done without loss of generality). More precisely, we assume that

(19.10.1) $$U = \bigcup_{j=1}^{r} \{M \in \mathfrak{M}(A): |\hat{x_k}(M)| < 1 \quad \text{for} \quad n_{j-1} < k \leqslant n_j\},$$

where $1 = n_0 < n_1 < \ldots < n_r$ and $x_2, \ldots, x_{n_r} \in A$.

Write $n = n_r$ and consider the following open subsets of C^n:

$$V = \bigcup_{j=1}^{r} \{z \in C^n: |z_k| < 1 \quad \text{for} \quad n_{j-1} < k \leqslant n_j\},$$

$$W = (C^n \setminus \bar{V}) \cup \{z: \operatorname{re} z_1 < 0\},$$

where z_k's are the coordinates of $z = (z_1, \ldots, z_n)$. We shall show that

$$\sigma(x_1, \ldots, x_n) \subset V \cup W, \tag{19.10.2}$$

where $x_2, \ldots, x_n$ are the elements occurring in (19.10.1) and $x_1 = x-1$, x being the element fixed at the outset. Indeed, if

$$M_0 \notin \{M \in \mathfrak{M}(A): |\hat{x_k}(M)| \leqslant 1 \quad \text{for} \quad n_{j-1} < k \leqslant n_j\},$$

then

$$(\hat{x_1}(M_0), \ldots, \hat{x_n}(M_0)) \notin \{z \in C^n: |z_k| \leqslant 1 \quad \text{for} \quad n_{j-1} < k \leqslant n_j\};$$

thus if $M_0 \notin \bar{U}$, then

$$(\hat{x_1}(M_0), \ldots, \hat{x_n}(M_0)) \notin \bar{V},$$

i.e.

$$(\hat{x_1}(M_0), \ldots, \hat{x_n}(M_0)) \in W. \tag{19.10.3}$$

If $M_0 \in \bar{U} \setminus E$, then $\operatorname{re} \hat{x_1}(M) < 0$ so that (19.10.3) again holds. Finally, if $M_0 \in U$, then clearly

$$(\hat{x_1}(M_0), \ldots, \hat{x_n}(M_0)) \in V$$

and inclusion (19.10.2) is proved.

According to Lemma 19.5 there exist elements $y_1, \ldots, y_p \in A$ such that

$$\sigma_{[x_1, \ldots, x_n, y_1, \ldots, y_p]}(x_1, \ldots, x_n) \subset V \cup W.$$

Then clearly

$$\sigma_{A_0}(x_1, \ldots, x_n, y_1, \ldots, y_p)$$
$$\subset \{z = (z_1, \ldots, z_{n+p}) \in C^{n+p}: (z_1, \ldots, z_n) \in V \cup W\} = Z,$$

where $A_0 = [x_1, \ldots, x_n, y_1, \ldots, y_p]$.

We restrict attention to the subalgebra $A_0 \subset A$. Similarly as in Lemma 19.4 (formula (19.4.5)) there exists a set D of the form (19.3.1) such that

$$\sigma_{A_0}(x_1, \ldots, x_n, y_1, \ldots, y_p) \subset D \subset U. \tag{19.10.4}$$

Putting

$$D_1 = \{z \in D: (z_1, \ldots, z_n) \in V\},$$

$$D_2 = \{z \in D: (z_1, \ldots, z_n) \in W\}$$

we obtain an open covering of D and we can apply the Oka theorem referred to in Definition 19.8. The Cousin data in question are now reduced to the single function $(\log z_1)/z_1$ defined in $D_1 \cap D_2$; this function is indeed well defined there since for $z = (z_1, \ldots, z_{n+p}) \in D_1 \cap D_2$ we have $\operatorname{re} z_1 < 0$ and so a branch of the logarithm can be fixed. By the Oka theorem this Cousin datum is solvable, i.e. there exist functions φ_1 and φ_2 defined in D_1 resp. D_2 and such that

$$\varphi_2(z)-\varphi_1(z) = \frac{\log z_1}{z_1} \quad \text{for} \quad z \in D_1 \cap D_2 .$$

Now let us put

$$\psi(z) = \begin{cases} z_1 e^{z_1 \varphi_1(z)} & \text{for} \quad z \in D_1, \\ e^{z_1 \varphi_2(z)} & \text{for} \quad z \in D_2. \end{cases}$$

This function is defined and analytic in D, thus, in view of (19.10.4) and Theorem 19.6, there exists an element $y \in A_0 \subset A$ such that

$$\psi(\hat{x_1}(M), \ldots, \hat{x_n}(M), \hat{y_1}(M), \ldots, \hat{y_p}(M)) = y\hat{\ }(M)$$

for every $M \in \mathfrak{M}(A_0)$ and a fortiori for every $M \in \mathfrak{M}(A)$.

Now consider the function $1/\psi$ restricted to the set $\sigma = \sigma_{A_0}(x_1, \ldots, x_n, y_1, \ldots$ $\ldots, y_p)$ and defined off the zeros of ψ. The function ψ is invertible on the set $\sigma \cap D_2$ so $\mathrm{re}(1/\psi)$ is bounded there. For $z \in \sigma \cap D_1$ we have

$$\frac{1}{\psi(z)} = \frac{1}{z_1} + \frac{e^{-z_1 \varphi_1(z)} - 1}{z_1}$$

and since the real part of the second summand is bounded and $\mathrm{re}(1/z_1) \leqslant 0$ for $z \in D_1$ so there exists a constant C such that

$$\mathrm{re}\frac{1}{\psi(z)} \leqslant C$$

for all $z \in \sigma$. It hence follows that the function

$$\mathrm{re}\frac{1}{y\hat{\ }(M)}$$

(defined off the zeros of $y\hat{\ }$) does not assume values in the half-plane $\{w: \mathrm{re}\, w > C\}$ and so the function $y\hat{\ }$ does not assume values in the disc $\{w: |w-1/C| < 1/C\}$.

Put $\varepsilon = 1/(2C)$ and $u = \varepsilon(\varepsilon e - y)^{-1} \in A$. The element u has the properties

(19.10.5) $\quad u\hat{\ }(M) = 1$ for $M \in E$ and $|u\hat{\ }(M)| < 1$ for $M \notin E$;

and this means that E is a peak set.

19.10.a. EXERCISE. Prove (19.10.5) in detail.

The above theorem implies another one, called the *local maximum modulus principle*.

19.11. THEOREM. *Let $A \in \mathbf{BC}_e$ and let U be an open subset of $\mathfrak{M}(A)$. Then*

$$\max_{M \in \bar{U}} |x\hat{\ }(M)| = \max_{M \in [U \cap \Gamma(A)] \cup \partial U} |x\hat{\ }(M)|$$

for every $x \in A$.

PROOF. Fix an element $x \in A$ and write

$$E = \{M_0 \in \bar{U} \colon |x\hat{}(M_0)| = \max_{M \in \bar{U}} |x\hat{}(M)|\}.$$

We have to show that $E \cap (\Gamma(A) \cup \partial U) \neq \emptyset$. So suppose that $E \cap \partial U = \emptyset$, i.e. $E \subset U$. Replacing, if necessary, x by some scalar multiple of x, we may assume that $\max_{M \in \bar{U}} |x\hat{}(M)| = 1$ and that $x\hat{}(M_0) = 1$ for some $M_0 \in E$. Putting

$$F = \{M \in U \colon x\hat{}(M) = 1\}$$

we obtain a closed subset $F \subset E \subset U$.

Write $u = \frac{1}{2}(e+x)$; then $u\hat{}(M) = 1$ for $M \in F$ and $|u\hat{}(M)| < 1$ for $M \in U \setminus F$ and so F is a local peak set. In virtue of the above theorem F is a peak set. Consequently F intersects the Shilov boundary $\Gamma(A)$ and thus also $E \cap \Gamma(A) \neq \emptyset$.

For further references to above topics cf. nicely written review article of Allan [3].

§ 20. DECOMPOSITION OF A COMMUTATIVE BANACH ALGEBRA INTO A DIRECT SUM OF IDEALS

As an application of Theorem 19.6 we give a proof of the decomposition theorem referred to in the title of this section.

20.1. DEFINITION. A Banach algebra A is said *to decompose into a direct sum of ideals* whenever there exist closed ideals $I_1, I_2 \subset A$ such that

(20.1.1) $$A = I_1 \oplus I_2,$$

i.e. every element $x \in A$ can be uniquely written as a sum

(20.1.2) $$x = x_1 + x_2$$

with $x_1 \in I_1$, $x_2 \in I_2$.

Let $A \in \mathbf{BC}_e$. Since the uniqueness of the representation (20.1.2) implies $I_1 \cap I_2 = \{0\}$, then $I_1 I_2 \subset I_1 \cap I_2 = \{0\}$. Applying (20.1.2) to the unit e we have $e = e_1 + e_2$, $e_i \in I_i$ and $e_1 e_2 = 0$, so that $e_1 + e_2 = e = e^2 = e_1^2 + e_2^2$, whence $e_i = e_i^2$, by the uniqueness of (20.1.2). The elements e_1 and e_2 are thus idempotents and a decomposition of A into a direct sum of ideals involves a decomposition of the unit into a sum of idempotents. And vice versa. For if $e = e_1 + e_2$ with e_1, e_2 idempotents then $e_1 + e_2 = e = e^2 = e_1^2 + e_2^2 + 2e_1 e_2 = e_1 + e_2 + 2e_1 e_2$ whence $e_1 e_2 = 0$. Putting $I_1 = e_1 A$, $I_2 = e_2 A$ we get the decomposition (20.1.1) into a direct sum of ideals; indeed, for any $x \in A$ we have $x = x(e_1 + e_2) = xe_1 + xe_2$, $xe_i \in I_i$, and if $x \in I_1 \cap I_2$, then $x = xe_1 = xe_2 = xe_1 e_2 = 0$.

Thus a decomposition of the unit into a sum of two proper idempotents (i.e. such that $0 \neq e_i \neq e$) corresponds to a decomposition of the algebra into a direct sum of proper ideals.

20.1.a. Exercise. Show that if e_1 is a proper idempotent of an algebra A, then so is the element $e_2 = e - e_1$.

20.1.b. Exercise. Show that if e_1 is a proper idempotent of an algebra A, then $e_1 A$ is a closed ideal in A.

20.1.c. Exercise. Show that if e_1 is an idempotent of an algebra A, then the ideal $e_1 A$ is an algebra with the unit e_1.

Now we prove the basic theorem on the decomposition of a Banach algebra into a direct sum of ideals.

20.2. Theorem. *An algebra $A \in \mathbf{BC}_e$ is decomposable into a direct sum of ideals* (20.1.1) *if and only if its maximal ideal space $\mathfrak{M}(A)$ is a union of two disjoint closed sets*

$$\mathfrak{M}(A) = F_1 \cup F_2.$$

More precisely: *if $A = I_1 \oplus I_2$, then $\mathfrak{M}(A) = \mathfrak{M}(I_1) \cup \mathfrak{M}(I_2)$* (each $I_i \in \mathbf{BC}_e$, see Exercise 20.1.c); conversely, *if $\mathfrak{M}(A) = F_1 \cup F_2$ with $F_i = \overline{F_i}$, $F_1 \cap F_2 = \emptyset$, then $A = I_1 \oplus I_2$ so that $\mathfrak{M}(I_i) = F_i$, $i = 1, 2$.*

Proof. Let $A = I_1 \oplus I_2$ and let $e = e_1 + e_2$ be the corresponding idempotent decomposition of the unit. Since $e_i^2 = e_i$, then for each $f \in \mathfrak{M}(A)$ either $f(e_i) = 0$ or $f(e_i) = 1$, $i = 1, 2$. Write $F_i = \{f \in \mathfrak{M}(A) : f(e_i) = 1\}$, $i = 1, 2$. Then clearly $\mathfrak{M}(A) = F_1 \cup F_2$ and each F_i is closed in $\mathfrak{M}(A)$. We shall show that $F_i = \mathfrak{M}(I_i)$. In fact, if $f \in F_i$, then the restriction $f|_{I_i}$ is a non-zero multiplicative linear functional on I_i and $f(e_i) = 1$. Distinct functionals have distinct restrictions. On the other hand, if $\varphi \in \mathfrak{M}(I_i)$ $(\varphi(e_i) = 1)$, then the functional f defined by $f(x) = \varphi(xe_i)$ is in $\mathfrak{M}(A)$ and $f|_{I_i} = \varphi$. So the restriction defines a one-to-one correspondence between F_i and $\mathfrak{M}(I_i)$, a homeomorphism, in fact, and the first assertion follows.

The other assertion is less trivial. Assume that $\mathfrak{M}(A) = F_1 \cup F_2$, $F_1 \cap F_2 = \emptyset$, $F_i = \overline{F_i}$, $i = 1, 2$. It suffices to construct an idempotent $e_1 \in A$ such that

$$\text{(20.2.1)} \qquad \hat{e_1}(M) = \begin{cases} 1 & \text{for} \quad M \in F_1, \\ 0 & \text{for} \quad M \in F_2. \end{cases}$$

For then $e_2 = e - e_1$ also is an idempotent (see Exercise 20.1.a) and the obtained idempotent decomposition of the unit will involve the asserted decomposition $A = I_1 \oplus I_2$ with $\mathfrak{M}(I_i) = F_i$, $i = 1, 2$, as can be easily seen.

The construction of the idempotent e_1 will involve more advanced tools such as analytic functions of several variables and the concept of the joint spectrum of any family of elements (Definition 17.14). We shall first show that there exist a finite system of elements $x_1, \ldots, x_n \in A$ and two open sets $U_1, U_2 \subset \mathbf{C}^n$ such that $U_1 \cap U_2 = \emptyset$ and

$$\{(\hat{x_1}(M), \ldots, \hat{x_n}(M)) \in \mathbf{C}^n : M \in F_i\} \subset U_i, \qquad i = 1, 2.$$

If the algebra A is finitely generated, then the generators $x_1, \ldots, x_n$ will do, since the joint spectrum $\sigma(x_1, \ldots, x_n)$ is then homeomorphic to the space

$\mathfrak{M}(A) = F_1 \cup F_2$ and the open sets U_1, U_2 clearly exist. We may thus assume that A is infinitely generated by some family of its elements x_λ, $\lambda \in \Lambda$: $A = [x_\lambda]_{\lambda \in \Lambda}$. Let σ denote the joint spectrum of this family of elements, a compact subset of the product $\boldsymbol{C}^\Lambda$. This spectrum is homeomorphic to $\mathfrak{M}(A)$ (Exercise 17.14.a), so that $\sigma = F_1 \cup F_2$, $F_1 \cap F_2 = \emptyset$, the compact sets F_i being regarded here as subsets of $\boldsymbol{C}^\Lambda$. Then there exist disjoint open sets $U_1, U_2 \subset \boldsymbol{C}^\Lambda$ such that $F_i \subset U_i$. Let L be any subset of the index set Λ and let

$$\boldsymbol{C}^L = \{\xi \in \boldsymbol{C}^\Lambda : \xi_\lambda = 0 \quad \text{for} \quad \lambda \notin L\}.$$

$\boldsymbol{C}^L$ is a subspace of $\boldsymbol{C}^\Lambda$. Let $\mathfrak{B}$ denote the family of sets $V_L \times \boldsymbol{C}^{\Lambda \setminus L}$ with L a finite subset of Λ, V_L open in $\boldsymbol{C}^L$; $\mathfrak{B}$ is a base for the topology of $\boldsymbol{C}^\Lambda$. Note that a finite union of members of $\mathfrak{B}$ is again in $\mathfrak{B}$.

The open set $U = U_1 \cup U_2$ is a union of some sets of the base $\mathfrak{B}$. The compact set σ, being covered by those sets, is covered by a finite number of them; in fact, is covered by a single element of $\mathfrak{B}$, in view of the preceding remark. We may thus assume that

$$U_i = V_P^{(i)} \times \boldsymbol{C}^{\Lambda \setminus P}, \qquad i = 1, 2,$$

where P is a finite subset of Λ, $P = \{\lambda_1, \ldots, \lambda_n\}$. Write $x_i = x_{\lambda_i}$. We are going to show that the joint spectrum $\sigma(x_1, \ldots, x_n)$ enjoys the desired properties. Consider the inclusion

$$\sigma \subset V_P^{(1)} \times \boldsymbol{C}^{\Lambda \setminus P} \cup V_P^{(2)} \times \boldsymbol{C}^{\Lambda \setminus P}. \tag{20.2.2}$$

The sets $V_P^{(i)}$ are disjoint, since the sets U_i were disjoint. Thus, after a projection of (20.2.2) onto the n-dimensional space $\boldsymbol{C}^P \subset \boldsymbol{C}^\Lambda$ we get $\sigma(x_1, \ldots, x_n) \subset V_P^{(1)} \cup V_P^{(2)}$, $V_P^{(1)} \cap V_P^{(2)} = \emptyset$ and the image of F_i lies in $V_P^{(i)}$. Now we shall restrict our attention to the space $\boldsymbol{C}^P$ and write U_1, U_2 instead of $V_P^{(1)}, V_P^{(2)}$. We thus have the desired situation:

$$\sigma(x_1, \ldots, x_n) \subset U_1 \cup U_2 \subset \boldsymbol{C}^n, \quad U_1 \cap U_2 = \emptyset,$$
$$\{(\hat{x_1}(M), \ldots, \hat{x_n}(M)) \in \boldsymbol{C}^n : M \in F_i\} \subset U_i.$$

Let φ be the function on $U = U_1 \cup U_2$ defined by $\varphi(\xi) = 1$ for $\xi \in U_1$, $\varphi(\xi) = 0$ for $\xi \in U_2$. The function φ is holomorphic in the neighbourhood of the spectrum $\sigma(x_1, \ldots, x_n)$, so, by Theorem 19.6, there exists an element $y \in A$ such that

$$y\hat{}(M) = \begin{cases} 1 & \text{for} \quad M \in F_1, \\ 0 & \text{for} \quad M \in F_2. \end{cases}$$

The element y need not be an idempotent, but certainly $y^2 - y \in \operatorname{rad} A$. The existence of the desired idempotent results now in virtue of the following lemma:

20.3. LEMMA. *If $x \in A$ and $x^2 - x \in \operatorname{rad} A$, then there exists an element $y \in A$ such that $y - x \in \operatorname{rad} A$ and $y^2 = y$.*

PROOF. Let $z_0 \in \operatorname{rad} A$ and write

(20.3.1) $$z_1 = -\tfrac{1}{2}\sum_{k=1}^{\infty}\binom{1/2}{k}(-4z_0)^k.$$

This is the power series expansion of the function $z_1 = \frac{1}{2}(1-\sqrt{1-4z_0})$. This series converges in A since it has a positive radius of convergence and $z_0 \in \operatorname{rad} A$ (cf. Definition 12.8). Hence $z_1^2 - z_1 + z_0 = 0$, as can be also verified by a direct computation. For any $z_0 \in \operatorname{rad} A$ the equation $x^2 - x + z_0 = 0$ has thus a solution (in x) in the radical. Now, let x be such that $x^2 - x \in \operatorname{rad} A$, according to the supposition. Equivalently, there is an element $q_0 \in \operatorname{rad} A$ such that

(20.3.2) $$x^2 - x + q_0 = 0.$$

We are looking for an element $u \in \operatorname{rad} A$ such that $(x+u)^2 = x+u$, or $u^2 - u(e-2x) + x^2 - x = 0$. This, by (20.3.2), may be written as

(20.3.3) $$u^2 - u(e-2x) - q_0 = 0.$$

The substitution $u = v(e-2x)$ transforms equation (20.3.3) into

(20.3.4) $$v^2(e-2x)^2 - v(e-2x)^2 - q_0 = 0.$$

But $(e-2x)^2 = e + 4(x^2 - x) = e - 4q_0$ is an invertible element of A since $q_0 \in \operatorname{rad} A$, thus equation (20.3.4) may be written as

$$v^2 - v - q_0(e-4q_0)^{-1} = 0.$$

This equation, as shown above, has a solution $v = q_1 \in \operatorname{rad} A$, $q_1^2 - q_1 - q_0(e-4q_0)^{-1} = 0$. Going back step by step we obtain the desired idempotent $y = x+u$, where $u = q_1(e-2x)$. This completes the proof of Lemma 20.3, thus also the proof of Theorem 20.2.

20.4. COROLLARY. *An algebra $A \in \mathbf{BC}_e$ can be decomposed into a direct sum of ideals $A = I_1 \oplus I_2 \oplus \dots \oplus I_n$ if and only if $\mathfrak{M}(A) = F_1 \cup F_2 \cup \dots \cup F_n$, with F_i pairwise disjoint closed sets; then $F_i = \mathfrak{M}(I_i)$, $i = 1, 2, \dots, n$.*

As a further application of the decomposition theorem we prove an extension of Lemma 20.3.

20.5. THEOREM. *Let $\varphi \not\equiv 0$ be a function holomorphic in an open set $U \subset \mathbf{C}$ and suppose that the equation*

$$\varphi(x) = q,$$

where $q \in \operatorname{rad} A$, has a solution $x \in A$ with $\sigma(x) \subset U$. Then there exists a solution $y \in A$ of the equation

$$\varphi(y) = 0$$

such that $x - y \in \operatorname{rad} A$.

PROOF. For any $f \in \mathfrak{M}$ we have $f(\varphi(x)) = \varphi(f(x)) = f(q) = 0$, and since $\sigma(x) \subset U$, the spectrum $\sigma(x)$ contains at most finitely many zeros of the function φ. Thus for any $f \in \mathfrak{M}$ one of the equalities $f(x) = \lambda_i$, $i = 1, \dots, n$, holds, where the λ_i's are the zeros of φ. Writing $F_i = \{M \in \mathfrak{M}: x\hat{}(M) = \lambda_i\}$ we get a decomposition of the space $\mathfrak{M}$ into a union of components $\mathfrak{M} = F_1 \cup \ldots \cup F_n$ which corresponds to a certain decomposition of A into a direct sum of ideals or a decomposition of the unit into a sum of idempotents: $e = e_1 + \ldots + e_n$. Moreover, $f(e_i) = \delta_{ij}$ for $f \in F_j$, $j = 1, \ldots, n$. Then $y = \sum_{i=1}^{n} \lambda_i e_i$ is the desired element, for $\varphi(y) = 0$ and $y - x \in \operatorname{rad} A$, since clearly $\varphi(xe_i) = e_i \varphi(x)$ for all $x \in A$.

20.5.a. Exercise. Let $A \in \mathbf{BC}_e$. Prove that $\exp x = \exp y$ implies $x - y = 2k\pi ie$ if and only if the space $\mathfrak{M}(A)$ is connected (the symbol $\exp x$ being defined by (16.2.1)).

Hint: show that the equation $\exp x = e$ has no non-zero solution in the radical and that it has solutions different from $2k\pi ie$ if and only if the space $\mathfrak{M}(A)$ is disconnected.

§ 21. LOCALLY ANALYTIC OPERATIONS. EVA KALLIN'S COUNTER-EXAMPLE

21.1. Definition. Let $A \in \mathbf{BC}_e$. We say that a function $f(M)$ defined on the space $\mathfrak{M} = \mathfrak{M}(A)$ *is obtained by means of a locally analytic operation* if for every point $M_0 \in \mathfrak{M}$ there exist a neighbourhood $U \subset \mathfrak{M}$ and elements $z_1, \ldots, z_n \in A$ such that

$$f(M) = \sum_{k_1, \ldots, k_n = 0}^{\infty} a_{k_1, \ldots, k_n} [\hat{z}_1(M) - \hat{z}_1(M_0)]^{k_1} \ldots [\hat{z}_n(M) - \hat{z}_n(M_0)]^{k_n}$$

for all $M \in U$, where the coefficients $a_{k_1, \ldots, k_n}$ are complex numbers and the series

$$\sum_{k_1, \ldots, k_n = 0}^{\infty} a_{k_1 \ldots k_n} \xi_1^{k_1} \ldots \xi_n^{k_n}$$

converges in some neighbourhood of the origin in C^n.

There existed, until recently, a conjecture stating that every such function is the Gelfand transform of some element of A. A consequence of this conjecture is another one, the so-called conjecture on functions locally belonging to A. A function $f(M)$ defined on $\mathfrak{M}$ is said *to belong to A locally* if every point $M_0 \in \mathfrak{M}$ has a neighbourhood in which f coincides with the Gelfand transform of some element of A (depending on M_0). The question whether every such function is (globally) the Gelfand transform of some element of A has been open for several years. The negative answer to this question, thus also to the conjecture concerning locally analytic operations, was given by Eva Kallin in the paper [1].

We give below the counter-example constructed by her.

21.2. Example. Put

$$\begin{aligned}
R &= \{\xi \in C^4: \xi_1 \xi_2 = 2,\ 1 \leqslant |\xi_1| \leqslant 2,\ \xi_3 = \xi_4 = 0\},\\
T_1 &= \{\xi \in C^4: \xi_1 \xi_2 = 2,\ |\xi_1| = 1,\ |\xi_3| \leqslant 1,\ \xi_4 = 0\},\\
T_2 &= \{\xi \in C^4: \xi_1 \xi_2 = 2,\ |\xi_1| = 2,\ |\xi_3| \leqslant 1,\ \xi_4 = \xi_3^2\},\\
X &= T_1 \cup R \cup T_2.
\end{aligned}$$

R is a surface homeomorphic to the ring $\{\xi \in C: 1 \leqslant |\xi| \leqslant 2\}$, T_1 and T_2 are (topologically) tori touching the inner, resp. the outer boundary circle of R. X is a connected subset of C^4, since $R \cap T_1 \neq \emptyset$ and $R \cap T_2 \neq \emptyset$. We shall construct a certain subalgebra of $C(X)$. Namely we put

$$x_1 = x_1(\xi) = \xi_1,$$
$$x_2 = x_2(\xi) = 2/\xi_1 \quad (= \xi_2 \text{ on } X),$$
$$x_3 = x_3(\xi) = \xi_3,$$
$$x_4 = x_4(\xi) = \begin{cases} 0 & \text{for} \quad \xi \in T_1 \cup R \\ \xi_3^2 & \text{for} \quad \xi \in T_2 \end{cases} \qquad (= \xi_4 \text{ on } X),$$
$$A = [x_1, x_2, x_3, x_4] \subset C(X).$$

We shall show that the function $g \in C(X)$ defined by

$$g(\xi) = \begin{cases} 0 & \text{for} \quad \xi \in T_1 \cup R, \\ \xi_3 & \text{for} \quad \xi \in T_2 \end{cases}$$

locally belongs to A but is not an element of A. Consider a measure μ whose support is contained in the boundary circles of R and which is orthogonal to all functions holomorphic in R. Write $\nu = (2\pi i)^{-1}\xi_3^{-2}d\xi_3$, a measure on $\{\xi \in X: |\xi_3| = 1\}$ and consider the product measure $\mu \times \nu$ defined on the set

$$B = \{\xi \in X: |\xi_1| = 1 \quad \text{or} \quad |\xi_1| = 2 \quad \text{and} \quad |\xi_3| = 1\}.$$

This set is homeomorphic (the homeomorphism being realized by the projections to the (ξ_1) and (ξ_3) planes) to the product

$$\{\xi_1 \in C: |\xi_1| = 1 \quad \text{or} \quad |\xi_1| = 2\} \times \{\xi_3 \in C: |\xi_3| = 1\}.$$

We shall show that the measure $\mu \times \nu$ is orthogonal to A. In fact, let p be any polynomial in four complex variables,

$$p(\xi_1, \ldots, \xi_4) = q(\xi_1, \xi_2, \xi_3) + \xi_4 r(\xi_1, \ldots, \xi_4),$$

where q and r are polynomials in three resp. four variables. We have

$$\int_B q d\nu = \frac{\partial q}{\partial \xi_3}\bigg|_{\xi_3 = 0}.$$

This is a function defined in R and holomorphic there (with respect to ξ_1, ξ_2), so

$$\int q d\mu \times \nu = 0.$$

Now we calculate the integral $\int_B \xi_4 r d\nu$. If $\xi \in R \cup T_1$, then $\xi_4 = 0$; if $\xi \in T_2$, then $\xi_4 = \xi_3^2$ and $\int \xi_4 r d\nu = (2\pi i)^{-1} \int_{|\xi_3|=1} r(\xi_1, \xi_2, \xi_3, \xi_3^2) d\xi_3 = 0$. Thus $\int_B p d\mu \times \nu = 0$ for any polynomial p, hence also for any element of A.

The measure μ may be specified as follows:

$$\mu = \frac{d\xi_1}{\xi_1} \quad \text{for} \quad |\xi_1| = 2 \quad \text{and} \quad \mu = -\frac{d\xi_1}{\xi_1} \quad \text{for} \quad |\xi_1| = 1.$$

Then we have

$$\int_B g d\mu \times \nu = \int_{B \cap T_2} g d\mu \times \nu = \int_{|\xi_1|=2} \int_{|\xi_3|=1} \xi_3 \frac{d\xi_1}{\xi_1} \cdot \frac{1}{2\pi i} \cdot \frac{d\xi_3}{\xi_3^2} = 2\pi i \neq 0,$$

whence it follows that $g \notin A$. To see that g belongs to A locally, observe that every point of X has a neighbourhood in which $g(\xi)$ is identically equal either to 0 or to ξ_3.

It remains to prove that $X = \mathfrak{M}(A)$. It will be sufficient to show that X is a polynomially convex subset of C^4 (cf. the second assertion of Theorem 17.10 and its proof). The set X will be shown to be precisely the intersection of the following sets:

$$\{\xi \in C^4 \colon \xi_1 \xi_2 = 2\},$$
$$\{\xi \in C^4 \colon |\xi_1| \leqslant 2\},$$
$$\{\xi \in C^4 \colon |\xi_2| \leqslant 2\},$$
$$\{\xi \in C^4 \colon |\xi_3| \leqslant 1\},$$
$$\{\xi \in C^4 \colon \xi_4(\xi_4 - \xi_3^2) = 0\},$$
$$\{\xi \in C^4 \colon |\xi_4 \xi_2^k| \leqslant 1\}, \quad k = 1, 2, \ldots,$$
$$\{\xi \in C^4 \colon |\xi_1^k(\xi_4 - \xi_3^2)| \leqslant 1\}, \quad k = 1, 2, \ldots$$

Let X_0 denote the intersection of these sets. Of course, X_0 is polynomially convex. It is not difficult to see that $X \subset X_0$. Write

$$Y = \{\xi \in C^4 \colon \xi_1 \xi_2 = 2, |\xi_1| \leqslant 2, |\xi_2| \leqslant 2, |\xi_3| \leqslant 1, \xi_4(\xi_4 - \xi_3^2) = 0\},$$

also a polynomially convex set: $X \subset X_0 \subset Y$. Write further

$$Y_1 = \{\xi \in Y \colon |\xi_1| < 2, \xi_4 = \xi_3^2 \neq 0\},$$
$$Y_2 = \{\xi \in Y \colon |\xi_1| > 1, \xi_4 = 0, \xi_4 - \xi_3^2 = -\xi_3^2 \neq 0\}.$$

Then $Y \diagdown X = Y_1 \cup Y_2$ (the verification is left to the reader). If $\xi \in Y_1$, then $|\xi_1| < 2$ and $|\xi_4| \neq 0$, so $|\xi_2| > 1$ and there exists such an integer k that $|\xi_3 \xi_2^k| > 1$, hence $\xi \notin X_0$. If $\xi \in Y_2$, then $|\xi_1| > 1$ and $|\xi_4 - \xi_3^2| \neq 0$, so $|\xi_1^k(\xi_4 - \xi_3^2)| > 1$ for some integer k and again $\xi \notin X_0$. It follows that $X_0 \subset Y \diagdown (Y_1 \cup Y_2) = X$, i.e. $X = X_0$. Thus X is a polynomially convex set and the proof is complete.

In the next section we shall be dealing with the class of the so called regular algebras. In this class both conjectures on functions locally belonging to the algebra and on locally analytic operations, are valid.

Note that the function g constructed in the above example is locally analytic on X but is not globally analytic there (for then it would belong to A by Theorem 19.6).

§ 22. REGULAR ALGEBRAS

22.1. DEFINITION. An algebra $A \in \mathbf{BC}_e$ is called a *regular algebra* if

(i) A is semi-simple,

(ii) for every closed subset $F \subset \mathfrak{M} = \mathfrak{M}(A)$ and every ideal $M_0 \in \mathfrak{M} \setminus F$ there exists an element $x \in A$ such that $x\hat{}(M) = 0$ for $M \in F$ and $x\hat{}(M_0) \neq 0$.

The following exercises provide examples for this definition.

22.1.a. EXERCISE. Prove that the algebra $A = C(\Omega)$, Ω compact Hausdorff, is regular.

22.1.b. EXERCISE. Prove that the algebra $\mathscr{A}$ of all functions holomorphic in the open disc and continuous in its closure is not regular.

In the theory of regular algebras an important role is played by the notion of the structure space $\mathfrak{M}_s$ (§ 11). We recall that the structure space of an algebra A is the set $\mathfrak{M}'$ of its maximal ideals equipped with the topology defined by the following closure operation. For a set $X \subset \mathfrak{M}'$ we put

$$\bar{X} = \{M \in \mathfrak{M}' \colon \bigcap X \subset M\},$$

where $\bigcap X = \bigcap\limits_{M \in X} M$. Writing

$$k(X) = \bigcap X$$

and

$$h(Y) = \{M \in \mathfrak{M}' \colon Y \subset M\}$$

for any subsets $X \subset \mathfrak{M}'$, $Y \subset A$, we obtain the equality

$$\bar{X} = h[k(X)].$$

22.2. THEOREM. *Let $A \in \mathbf{BC}_e$ be a regular algebra. Then for any closed subset $F \subset \mathfrak{M}$ (1) there exists a closed ideal $I \subset A$ such that $h(I) = F$.*

PROOF. We set $I = \{x \in A \colon x\hat{}(M) = 0 \text{ for } M \in F\}$. Clearly I is a closed ideal in A. If $M_0 \in \mathfrak{M} \setminus F$, then there exists an element $x_0 \in I$ such that $\hat{x_0}(M_0) \neq 0$, whence $x_0 \notin M_0$ and $M_0 \notin h(I)$. Hence $h(I) \subset F$. On the other hand, if $M \subset F$ then $I \subset M$ so that $M \in h(I)$. Thus we get the inclusion $F \subset h(I)$ and consequently $h(I) = F$.

The ideal I constructed in the proof of Theorem 22.2 will be denoted by the symbol I_F.

The following theorem shows that regular algebras are also "normal".

22.3. THEOREM. *Let $A \in \mathbf{BC}_e$ be regular and let F_1, F_2 be disjoint closed non-void subsets of $\mathfrak{M}$. Then there exists an element $x_0 \in A$ such that*

$$\hat{x_0}(M) = \begin{cases} 1 & \text{for} \quad M \in F_1, \\ 0 & \text{for} \quad M \in F_2. \end{cases}$$

---

(1) Closure of a set $F \subset \mathfrak{M}$ means here closure with respect to the Gelfand topology. If we meant closure with respect to the structure topology, we would speak of a subset $F \subset \mathfrak{M}_s$

PROOF. Consider the quotient algebra $A_1 = A/I_{F_1}$ and the image $I' = \varphi(I_{F_2}) \subset A_1$ of the ideal I_{F_2} under the natural homomorphism $\varphi: A \to A_1$. The set I', being a homomorphic image of an ideal, either is a proper ideal in A_1 or is the whole algebra A_1. Suppose that I' is a proper ideal in A_1. Then it is contained in some maximal ideal $M_0' \in \mathfrak{M}(A_1)$ and the counterimage $\varphi^{-1}(M_0') \subset A$ is a maximal ideal in A containing both I_{F_1} and I_{F_2}. This means that $M_0 \in F_1 \cap F_2$, contrary to the assumption that F_1 and F_2 are disjoint. Therefore $I' = A_1$ and there exists an element $x_0 \in I_{F_2}$ such that $\varphi(x_0)$ is the unit of A_1. Since $\mathfrak{M}(A_1) = F_1$, then $\hat{x_0}(M) = 1$ for $M \in F_1$. On the other hand, $x_0 \in I_{F_2}$, i.e. $\hat{x_0}(M) = 0$ for $M \in F_2$ and thus x_0 is the desired element.

22.3.a. EXERCISE. Prove that the algebra A_1 constructed in the proof of Theorem 22.3 is a regular algebra.

REMARK. Note that in the proof of Theorem 22.3 any ideal I such that $h(I) = F_2$ could be taken in place of I_{F_2}. As a corollary we obtain the following theorem.

22.4. THEOREM. *Let $A \in \mathbf{BC}_e$ be a regular algebra. If F_1 and F_2 are disjoint closed non-void subsets of the space $\mathfrak{M}(A)$ and if I is an ideal in A such that $h(I) = F_2$, then there exists an element $x_0 \in I$ with the property*

$$\hat{x_0}(M) = \begin{cases} 1 & \textit{for} \quad M \in F_1, \\ 0 & \textit{for} \quad M \in F_2. \end{cases}$$

The following theorem characterizes regular algebras by means of the concept of the structure space.

22.5. THEOREM. *An algebra $A \in \mathbf{BC}_e$ is regular if and only if the identity map of $\mathfrak{M}$ onto $\mathfrak{M}_s$ is a homeomorphism.*

PROOF. Suppose that A is regular and let F be a closed subset of $\mathfrak{M}(A)$. We are going to show that then F is also a closed subset of $\mathfrak{M}_s(A)$, i.e. $F = h(k(F))$. We have

$$k(F) = \bigcap F = \{x \in A : x\hat{}(M) = 0 \quad \text{for} \quad M \in F\} = I_F$$

and, by Theorem 22.2, $h(I_F) = F$, so $F = \bar{F}$ in $\mathfrak{M}_s$.

Conversely, we shall show that any closed subset of $\mathfrak{M}_s$ is also closed in $\mathfrak{M}$. Closed subsets of $\mathfrak{M}_s$ are precisely the sets $h(k(F))$, so it suffices to show that for a given ideal I the set $h(I)$ is a closed subset of $\mathfrak{M}$. Write $F_x = \{M \in \mathfrak{M}: x \in M\}$. F_x is a closed subset of $\mathfrak{M}$, for if $M_0 \notin F_x$, then $x(M_0) = \alpha \neq 0$ and the neighbourhood $V(M_0, x, \frac{1}{2}|\alpha|)$ is disjoint with F_x; so the set $h(I)$ is closed in view of the equality $h(I) = \bigcap_{x \in I} F_x$. We see that the family of closed subsets of $\mathfrak{M}$ coincides with the family of closed subsets of $\mathfrak{M}_s$ and thus the two topologies coincide.

Now suppose that the algebra A is not regular. Then there exists a closed set $F_0 \subset \mathfrak{M}$ and an ideal $M_0 \in \mathfrak{M} \setminus F_0$ such that every function $x^{\wedge}(M)$ which is zero on the set F_0 vanishes also at M_0. So we have

$$M_0 \in h(k(F_0)) = \{M \in \mathfrak{M}: \bigcap F_0 \subset M\},$$

i.e. M_0 is in the closure of F_0 in $\mathfrak{M}_s$, whence it follows that the set F_0 is not closed in $\mathfrak{M}_s$ and the two topologies are not equivalent.

In Theorems 22.7-22.12 A denotes a regular commutative Banach algebra with unit.

22.6. Corollary. *All theorems of this section remain valid if the space $\mathfrak{M}$ is replaced by $\mathfrak{M}_s$.*

22.7. Theorem. *The Shilov boundary of A is the whole space $\mathfrak{M}(A)$.*

Proof. Let $M_0 \in \mathfrak{M}$. In order to prove that $M_0 \in \Gamma(A)$ it suffices to show that for any neighbourhood U of M_0 there exists an element $x \in A$ such that

$$\sup_{M \in \mathfrak{M} \setminus U} |x^{\wedge}(M)| < \sup_{M \in U} |x^{\wedge}(M)|$$

(Theorem 15.3); and such an element can be obtained by putting in Theorem 22.3 $F_1 = \{M_0\}$ and $F_2 = \mathfrak{M} \setminus U$.

22.7.a. Exercise. Find out whether the converse statement is true (give a proof or a counterexample)

The following *theorem on the partition of unity* will be applied in the sequel to proofs of theorems on functions locally belonging to an algebra and on locally analytic functions.

22.8. Theorem. *For any open covering $\mathfrak{M}(A) = U_1 \cup \ldots \cup U_n$ there exist elements $h_1, \ldots, h_n \in A$ such that*

(22.8.1) $$h_i^{\wedge}(M) = 0 \quad \textit{for} \quad M \in \mathfrak{M} \setminus U_i, \quad i = 1, \ldots, n,$$

(22.8.2) $$\sum_{i=1}^{n} h_i^{\wedge}(M) = 1 \quad \textit{for} \quad M \in \mathfrak{M}.$$

Proof. The proof will be by induction on n. Let $n = 2$. The sets $F_i = \mathfrak{M} \setminus U_i$, $i = 1, 2$, are closed and disjoint. According to Theorem 22.3 there exists an element $h_1 \in A$ such that

$$h_1^{\wedge}(M) = \begin{cases} 1 & \text{for} \quad M \in F_1, \\ 0 & \text{for} \quad M \in F_2. \end{cases}$$

Putting $h_2 = e - h_1$ we obtain the asserted partition. Now assume that the assertion is true for any covering consisting of at most $(n-1)$ open sets and let $U_1, \ldots, U_n$ be an open cover of $\mathfrak{M}$. Write $F = \mathfrak{M} \setminus U_n$. We have $F \subset U_1 \cup \ldots \cup U_{n-1} = V$. Since $\mathfrak{M}$ is a normal space, so there exists an open set $U \subset \mathfrak{M}$ such that $F \subset U \subset \overline{U} \subset V$. Consider the algebra $A_1 = A/I_{\overline{U}}$. We know (Theorem

11.3) that $\mathfrak{M}(A_1) = \bar{U}$. Since A_1 is a regular algebra (Exercise 22.3.a) and $\mathfrak{M}(A_1)$ is covered by open sets $U_i \cap \bar{U}$, $i = 1, \ldots, n-1$ (every functional $f \in \mathfrak{M}$ which vanishes on $I_{\bar{U}}$ may be regarded as a functional defined in A_1), then by the induction hypothesis, there exist elements $h_1', \ldots, h_{n-1}' \in A_1$ satisfying the analogues of conditions (22.8.1) and (22.8.2) in the algebra A_1. Let $h_1'', \ldots, h_{n-1}'' \in A$ denote any counter-images of those elements under the natural homomorphism $A \to A_1$. Since $U \cup U_n = \mathfrak{M}(A)$, so there exist in A elements h_0 and h_n with the properties $h_0^\wedge(M) = 0$ for $M \in U$, $h_n^\wedge(M) = 0$ for $M \in U_n$ and $h_0^\wedge(M) + h_n^\wedge(M) \equiv 1$ on $\mathfrak{M}(A)$. Write $h_i = h_0 h_i''$, $i = 1, \ldots, n-1$. We have $h_i^\wedge(M) = 0$ for $M \notin U_i$, $i = 1, \ldots, n$, since $h_i''(M) = 0$ for $M \in \bar{U} \setminus U_i$ and $h_0^\wedge(M) = 0$ for $M \notin U$. We then have

$$\sum_{i=1}^{n} h_i^\wedge(M) = \Big[\sum_{i=1}^{n-1} (h_i'')^\wedge(M) \Big] h_0^\wedge(M) + h_n^\wedge(M) = h_0^\wedge(M) + h_n^\wedge(M) \equiv 1.$$

22.9. Theorem. *Every function which locally belongs to A is the Gelfand transform of some element of this algebra.*

Proof. Suppose that a function $f(M)$ locally belongs to A. This means that for any $M_0 \in \mathfrak{M}$ there exist a neighbourhood U_{M_0} of this point and an element $x_{M_0} \in A$ such that $f(M) = x_{M_0}^\wedge(M)$ for all $M \in U_{M_0}$. Since $\mathfrak{M} \subset \bigcup_{M \in \mathfrak{M}} U_M$, then by the compactness of $\mathfrak{M}$, there exist points $M_1, \ldots, M_n \in \mathfrak{M}$ such that $\mathfrak{M} \subset \bigcup_{i=1}^{n} U_{M_i}$ a nd $f(M) = x_{M_i}^\wedge(M)$ for $M \in U_{M_i}$, $i = 1, \ldots, n$. Let $h_1, \ldots, h_n$ denote elements of A satisfying conditions (22.8.1) and (22.8.2) of the preceding theorem with $U_i = U_{M_i}$. We have

$$f(M) = \sum_{i=1}^{n} f(M) h_i^\wedge(M) = \sum_{i=1}^{n} x_i^\wedge(M) h_i^\wedge(M),$$

whence, putting $y = \sum_{i=1}^{n} x_i h_i$, we get $y^\wedge(M) = f(M)$ for all $M \in \mathfrak{M}$.

22.10. Theorem. *Let $f(M)$ be a function on $\mathfrak{M}(A)$ which is obtained by means of a locally analytic operation. Then $f(M)$ is the Gelfand transform of an element $x \in A$.*

Proof. For any element $M_0 \in \mathfrak{M}$ there exist a neighbourhood U_0 of it and such elements $z_1, \ldots, z_n \in A$ that

$$f(M) = \sum_{i_1, \ldots, i_n = 0}^{\infty} a_{i_1 \ldots i_n} (z_1^\wedge(M) - z_1^\wedge(M_0))^{i_1} \ldots (z_n^\wedge(M) - z_n^\wedge(M_0))^{i_n}$$

for $M \in U_0$, whereby the power series

$$\Phi(\xi) = \sum_{i_1, \ldots, i_n = 0}^{\infty} a_{i_1 \ldots i_n} \xi_1^{i_1} \ldots \xi_n^{i_n}$$

converges in some neighbourhood of zero $Q_\varepsilon = \{\xi \in C^n\colon |\xi_i| < \varepsilon,\ i = 1, \ldots, n\}$. Let $U_1(M_0) = U_0 \cap V(M_0; z_1, \ldots, z_n; \frac{1}{2}\varepsilon)$ and let us choose such neighbourhoods V_0 and W_0 of M_0 that $\overline{W}_0 \subset V_0 \subset \overline{V}_0 \subset U_1(M_0)$. In virtue of Theorem 22.3 there exists an element $h \in A$ such that

$$h^\wedge(M) = \begin{cases} 1 & \text{for} \quad M \in W_0, \\ 0 & \text{for} \quad M \notin V_0. \end{cases}$$

Since the joint spectrum $\sigma[(z_1 - \hat{z_1}(M_0)e)h_1, \ldots, (z_n - \hat{z_n}(M_0)e)h_n]$ is contained in Q_ε, then, by Theorem 19.1, there exists an element $y_0 = \Phi[(z_1 - \hat{z_1}(M_0)e)h_1, \ldots, (z_n - \hat{z_n}(M_0)e)h_n] \in A$. Moreover, we have $\hat{y_0}(M) = f(M)$ for $M \in W_0$ and since the point M_0 can be arbitrarily chosen, the function $f(M)$ locally belongs to A. Thus, because of the preceding theorem, there exists an element $x \in A$ such that $f(M) = x^\wedge(M)$ for all $M \in \mathfrak{M}$.

We shall now be concerned with the structure of ideals in regular algebras.

22.11. THEOREM. *Let F be a closed subset of $\mathfrak{M}(A)$. Then there exists the smallest ideal in the class of all ideals fulfilling the relation $h(I) = F$. This ideal is given by the following formula*:

$I_F^0 = \{x \in A$*: there exists an open set $U \supset F$ such that $x^\wedge(M) = 0$ for $M \in U$*$\}$.

PROOF. We first show that $h(I_F^0) = F$. Clearly we have $F \subset h(I_F^0)$. On the other hand, if $M_0 \notin F$, then there exists an open set $U \subset \mathfrak{M}$ such that $F \subset U$ and $M_0 \notin \overline{U}$. Thus there exists an element x_0 such that $\hat{x_0}(M_0) \neq 0$ and $\hat{x_0}(M) = 0$ for $M \in \overline{U}$. So $x_0 \in I_F^0$ and $x_0 \notin M_0$ whence $M_0 \notin h(I_F^0)$ and $F = h(I_F^0)$ follows. Now let I be any ideal in A such that $h(I) = F$. We shall show that $I_F^0 \subset I$ so that I_F^0 is indeed the smallest ideal with this property. Let $x \in I_F^0$; then there exists an open set $U \subset \mathfrak{M}$ such that $F \subset U$ and $x^\wedge(M) = 0$ for $M \in U$. Since the sets F and $\mathfrak{M} \setminus U$ are closed and disjoint, then by Theorem 22.4, there exists an element $z \in I$ with

$$z^\wedge(M) = \begin{cases} 1 & \text{for} \quad M \in M \setminus U, \\ 0 & \text{for} \quad M \in F. \end{cases}$$

Thus for every $M \in \mathfrak{M}$ we have $x^\wedge(M) - x^\wedge(M)z^\wedge(M) = 0$ and since A is semi-simple, so $x - xz = 0$ and $x = xz \in I$, which completes the proof of the inclusion $I_F^0 \subset I$.

22.12. THEOREM. *The ideal $\overline{I_F^0}$ is the smallest ideal in the class of all closed ideals I fulfilling the relation $h(I) = F$. Moreover,*

(22.12.1) $\overline{I_F^0} = \{x \in A$*: there exist a sequence $(x_n) \subset A$ with* $\lim x_n = 0$ *and a sequence of open sets (U_n) such that $U_n \supset F$ and $x^\wedge(M) = \hat{x_n}(M)$ for $M \in U_n$, $n = 1, 2, \ldots$*$\}$.

PROOF. Since $h(I) = h(\overline{I})$ for every ideal I, then $h(\overline{I_F^0}) = F$. We have, because of the preceding theorem, $I_F^0 \subset I$ for any closed ideal I fulfilling the condition

$h(I) = F$, thus also $\overline{I_F^0} \subset I$ and so $\overline{I_F^0}$ is in fact the smallest ideal in the class. Now, let $x \in \overline{I_F^0}$, i.e. $x = \lim y_n$, where $y_n \in I_F^0$. For every n there exists an open set $U_n \subset \mathfrak{M}$ such that $F \subset U_n$ and $y_n^\wedge(M) = 0$ for $M \in U_n$. Writing $x_n = x - y_n$ we have $\lim x_n = 0$ and $x_n^\wedge(M) = x^\wedge(M)$ for $M \in U_n$. Thus $\overline{I_F^0}$ is contained in the set on the right-hand side of (22.12.1). Conversely, if x belongs to this set, then $y_n = x - x_n$ converges to x and $y_n \in I_F^0$, so $x \in \overline{I_F^0}$ and the converse inclusion also holds.

Applying Theorem 22.12 to the algebra $C(X)$ we obtain the following theorem.

22.13. THEOREM. *If X is a compact Hausdorff space, then every closed ideal of the algebra $C(X)$ is of the form I_F.*

PROOF. Let I be a closed ideal in $C(X)$. Write $F = h(I)$. Obviously we have $\overline{I_F^0} \subset I \subset I_F$, so it suffices to prove that $I_F \subset \overline{I_F^0}$. Let $f \in I_F$. Fix an $\varepsilon > 0$ and consider the sets

$$F_1^\varepsilon = \{M \in X: |f(M)| \leqslant \varepsilon\}, \quad F_2^\varepsilon = \{M \in X: |f(M)| \geqslant 2\varepsilon\}.$$

The sets F_1^ε and F_2^ε are disjoint closed subsets of the compact space X, so, by the Urysohn theorem, there exists a function $g_\varepsilon \in C(X)$ such that

$$g_\varepsilon(M) = \begin{cases} 1 & \text{for} \quad M \in F_1^\varepsilon, \\ 0 & \text{for} \quad M \in F_2^\varepsilon, \end{cases}$$

and $0 \leqslant g_\varepsilon(M) \leqslant 1$. So we have $||g_\varepsilon f|| \leqslant 2\varepsilon$, and since $F \subset \operatorname{int} F_1^\varepsilon$ and $(g_\varepsilon f)(M) = f(M)$ for $M \in F_1^\varepsilon$, then $f \in \overline{I_F^0}$ by (22.12.1). Thus $I_F \subset \overline{I_F^0}$.

22.14. DEFINITION. A Banach algebra A is said to have *spectral synthesis* if every closed ideal of this algebra is precisely the intersection of all maximal ideals containing this ideal.

Since clearly $I_F = \bigcap_{M \in F} M = \bigcap_{I_F \in M} M$, it follows from Theorem 22.13 that all algebras $C(X)$ have spectral synthesis.

A simple example of an algebra without spectral synthesis is the algebra $\mathscr{A}$ of all functions holomorphic in the open disc $\{\lambda \in \mathbf{C}: |\lambda| < 1\}$, continuous on its boundary [1]. Consider the closed ideal

$$I = \{x \in \mathscr{A}: x(0) = x'(0) = 0\}.$$

This ideal is contained in exactly one maximal ideal $M = \{x \in \mathscr{A}: x(0) = 0\}$ and is not the whole ideal M. In the sequel we shall be concerned with ideals of this kind.

---

[1] The algebra $\mathscr{A}$ is not regular. There exist also regular algebras without spectral synthesis, e.g. $L_1(-\infty, \infty)$ with adjoined unit is such an algebra. In general: the group algebra $L_1(G)$ of a locally compact abelian group G has spectral synthesis if and only if the group G is compact (*Malliavin's theorem*).

22.15. DEFINITION. An ideal I of an algebra A is called a *primary ideal* if it is contained in exactly one maximal ideal. The algebra A is called *primary* if it has only one maximal ideal.

Thus an ideal I is primary if and only if the algebra A/I is primary. From Theorem 22.12 it follows that every maximal ideal M of a regular Banach algebra A contains exactly one minimal closed primary ideal, namely the ideal $\overline{I^0_{\{M\}}}$. If the algebra A is not regular, then the minimal closed primary ideal need not exist. For instance, in the algebra $\mathscr{A}$ the sequence

$$I_n = \{x \in \mathscr{A} : x(0) = x'(0) = \dots = x^{(n)}(0) = 0\}$$

is a descending sequence of primary ideals and $\bigcap I_n = \{0\}$, thus a minimal primary ideal does not exist.

22.15.a. EXERCISE. Prove that the Banach algebra $C^{(2)}[0, 1]$ is regular. Find the general form of a closed primary ideal in this algebra.

22.15.b. EXERCISE. Prove that a quotient algebra of a semi-simple algebra need not be semi-simple.

HINT: Consider the quotient algebra A/I, where I is a primary ideal but not a maximal ideal.

We mention here yet another application of the theory of regular algebras, the *Wiener Tauberian theorem*:

22.16. THEOREM. *Let $x_0(t) \in L_1(-\infty, \infty)$. Assume that $x_0(s) \neq 0$, where*

$$x^\wedge(s) = \int_{-\infty}^{\infty} x(t) e^{its} dt.$$

Let $f(t)$ be a bounded measurable complex-valued function defined on the real line and such that the limit

$$\lim_{\tau\to\infty} \int_{-\infty}^{\infty} x_0(t-\tau) f(t) dt = l \int_{-\infty}^{\infty} x_0(t) dt$$

exists for some constant l. Then the analogous limit exists and the equality

$$\lim_{\tau\to\infty} \int_{-\infty}^{\infty} x(t-\tau) f(t)\, dt = l \int_{-\infty}^{\infty} x(t) dt \tag{22.16.1}$$

holds for any $x \in L_1(-\infty, \infty)$.

OUTLINE OF THE PROOF. We consider the algebra $L_1(-\infty, \infty)$ with the convolution multiplication (Example 3.3). Each multiplicative linear functional in this algebra has the form $f(x) = x^\wedge(t)$ for some fixed real t (see 9.2). If $V = L_1 \oplus \{\lambda e\}$, then L_1 is a maximal ideal in V and $\mathfrak{M}(V) = \boldsymbol{R} \cup \{M_\infty\}$, where $\boldsymbol{R}$ is the real line and $M_\infty = L_1$ is "the point at infinity" which compactifies the real line $\boldsymbol{R}$. The algebra V is regular as follows from the fact that for any compact subset F of $\boldsymbol{R}$ and for any open set $U \subset \boldsymbol{R}$, $F \subset U$, there exists an element

$x \in L_1(-\infty, \infty)$ with the properties $x\hat{}(s) = 1$ for $s \in F$ and $x\hat{}(s) = 0$ for $s \notin U$ (Naimark [1]). It can be proved that the set of those elements of $L_1(-\infty, \infty)$ whose Fourier transforms have compact supports is dense in $L_1(-\infty, \infty)$. Since the set of those elements is just the ideal $L^0_{\{M_\infty\}}$, the only closed primary ideal contained in M_∞ is the ideal $M_\infty = L_1(-\infty, \infty)$ itself.

Let I denote the set of those elements of $L_1 \subset V$ which fulfil condition (22.16.1), or, after a change of variables, the condition

$$\text{(22.16.2)} \qquad \lim_{\tau\to\infty} \int_{-\infty}^{\infty} f(t+\tau)\,x(\tau)\,d\tau = l \int_{-\infty}^{\infty} x(t)\,dt.$$

We first show that I is a closed subspace of L_1. Observe that the norms of the functionals $\langle f_t, x\rangle = \int_{-\infty}^{\infty} f(t+\tau)x(\tau)d\tau$ have a common bound

$$\|f_t\| = \operatorname*{ess\,sup}_{\tau} |f(t+\tau)| = \operatorname*{ess\,sup}_{\tau} |f(\tau)| = \|f\|_\infty < \infty.$$

If $x_0 \in \bar{I}$, then $x_0 = \lim_{t\to\infty} x_n$, $x_n \in I$. We write relation (22.16.2) as $\lim\langle f_t, x\rangle = \langle l, x\rangle$ and thus obtain

$$|\langle f_t, x_0\rangle - \langle l, x_0\rangle| \leqslant |\langle f_t, x_0 - x_n\rangle| + |\langle f_t, x_n\rangle - \langle l, x_n\rangle| + |\langle l, x_n - x_0\rangle|$$
$$\leqslant (\|f\| + |l|)\,\|x_n - x_0\| + |\langle f_t, x_n\rangle - \langle l, x_n\rangle|.$$

Fix an $\varepsilon > 0$ and choose an integer N such that

$$\|x_N - x_0\| < \frac{\varepsilon}{2(\|f\| + |l|)},$$

and a number T such that $|\langle f_t, x_N\rangle - \langle l, x_N\rangle| < \frac{1}{2}\varepsilon$ holds for $t > T$. We get

$$|\langle f_t, x_0\rangle - \langle l, x_0\rangle| < \varepsilon$$

for $t > T$. Thus $x_0 \in I$ and the subspace I is closed in L_1. It is easy to conclude hence that I is an ideal in L_1 and thus also in V. Indeed we have

$$\int_{-\infty}^{\infty} f(t+\tau)\,x(\tau+p)\,d\tau = \int_{-\infty}^{\infty} f(t-p+\tau)\,x(\tau)\,d\tau,$$

so that I contains together with x all translates $x_t(\tau) = x(t+\tau)$; consequently, I is an ideal since the convolution $x * y = \int_{-\infty}^{\infty} x(t-\tau)\,y(\tau)\,d\tau$ can be approximated in L_1 by linear combinations of translates of x. Now, since $x_0 \in I$ and $\hat{x_0}(s) \neq 0$, then M_∞ is the only maximal ideal containing I. Thus I is a closed primary ideal contained in M_∞ and hence $I = M_\infty$, which means that (22.16.1) holds for all elements x of L_1.

Theorem 22.16 is often formulated as follows: *Every proper closed maximal ideal of the algebra L_1 is contained in a maximal modular ideal* (see Definition 6.9).

The same Wiener Tauberian theorem may be formulated in yet other terms. A closed set $F \subset \mathfrak{M}(A)$ is said to be a *set of spectral synthesis* or *to have spectral synthesis* if $I_F = \overline{I_F^0}$, i.e. if there exists exactly one closed ideal $I \subset A$ such that $h(I) = F$. The formulation referred to above reads as follows: *the empty set has spectral synthesis in the algebra* $L_1(-\infty, \infty)$.

It can also be seen from the above outline of the proof that the group algebra $L_1(-\infty, \infty)$ is generated by all translates of any function whose Fourier transform vanishes nowhere.

The notion of a regular algebra may be generalized also to non-commutative algebras with or without unit in such a way that most of the theorems of this section remain valid. Readers interested in this theory are referred to Rickart's book [1], § 7.

COMMENTS ON CHAPTER III

C.16. Results given here are true for p-normed algebras. Theorem 16.2 can be obtained without use of integrals e.g. by the use of power series (and was so obtained originally in Żelazko [5]).

Later on it was shown, that the integral methods can be also used in theory of p-normed algebras since the integrals involved here do exist (this was obtained independently by D. Przeworska-Rolewicz and S. Rolewicz in [1] and by B. Gramsch in [1], who, later on, in a series of papers [2]-[4] simplified many earlier proofs concerning p-normed algebras). Recently Allan, Daves and McClure [1] gave another method, reducing the problem to that of a Banach algebra.

The results given here are also true for m-convex algebras, and what is of some interest, Theorem 16.2 is here true, even if $\sigma(x) = \Omega$ (if it happens that $\sigma(x)$ it itself an open set). They are also true for LC-algebras if we take instead of $\sigma(x)$ — the closure $\overline{\Sigma_A(x)}$ of the extended spectrum (in the Riemann sphere). It turns out, however, that very often $\Sigma_A(x)$ is dense in the Riemann sphere and in this case the only surely operating functions are polynomials.

The problems concerned with entire functions defined by means of power series were already discussed in C.7.

The functional calculus makes sense also in algebras with a boundedness structure (cf. remarks at the end of C.10).

C.17. The results given here are true for p-normed algebras. Concerning LC-algebras, we discussed some topics of this section in C.10.

Theorem 17.8 is true if we replace a Banach algebra by an m-convex B_0-algebra (cf. Arens [5]).

Since it has important consequences (e.g. Theorem C.8.1 is an immediate consequence of this theorem), it would be interesting to know whether it is true for all (complete) m-convex algebras. We quote now an equivalent version of this theorem.

C.17.1. THEOREM. *Let A be a commutative m-convex B_0-algebra with unit e. Suppose that $F \in \mathfrak{M}^{\#}(A)$ and $x_1, x_2, \ldots, x_n \in A$. Then there exists a functional $f \in \mathfrak{M}(A)$ such that $F(x_i) = f(x_i)$ for $i = 1, 2, \ldots, n$.*

It was mentioned in C.10 that the joint spectrum of generators of an m-convex B_0-algebra need not be polynomially convex, so Theorem 17.10 fails in this case.

C.18. The results given here are true for p-normed algebras. Theorem 18.2 is true for m-convex algebras (provided the functional in question is continuous, which automatically holds in the case of a Banach algebra. If we drop the assumption of continuity, then probably the theorem fails, but we know of no such example), but fails for non-m-convex B_0-algebras (cf. Kahane, Żelazko [1]).

We pose two conjectures ([1]), each implying the existence of suitable counter-example. Put $S = \{e^{\varphi}: \varphi(0) = 0, \varphi$ an entire function of one complex variable$\}$, our conjectures read now:

1° S is a linearly independent set.

2° Finite linear combinations of elements of S do not exhaust all entire functions.

C.19. The results given here are true for commutative p-normed algebras, but it is necessary to use the integral methods mentioned in C.16. Theorems 19.6 and 19.7 are true for commutative m-convex algebras (cf. Allan [1], Arens [7], Bonnard [1]).

For a commutative LC-algebra A Theorem 19.7 is also true under additional assumptions about elements $x_1, \ldots, x_n \in A$ (cf. Waelbroeck [1]-[8]).

([1]) The positive answer to both conjectures follows from a result in Borel [1] (added in the proof).

CHAPTER IV

Algebras with an Involution

The theory of algebras with an involution is a branch of the theory of Banach algebras and certainly one of the most important for applications. Here we shall restrict ourselves to an outline of the basic facts of the theory of the so-called **B*-*algebras*** (= **C*-*algebras***) and the theory of the representations of general algebras with an involution.

§ 23. INTRODUCTORY REMARKS

In this section no commutativity or existence of a unit in the algebras in question is assumed.

23.1. DEFINITION. A mapping $x \to x^*$ of a Banach algebra A onto itself is called an *involution* if it satisfies the following conditions:

(i) $(x^*)^* = x$,

(ii) $(x+y)^* = x^*+y^*$,

(iii) $(xy)^* = y^*x^*$,

(iv) $(\alpha x)^* = \bar{\alpha}x^*$,

where $x, y \in A$ and α is a complex scalar. Here we give a few examples of algebras with an involution. The verification of conditions (i)–(iv) is left to the reader as an exercise.

23.2. $A = C(\Omega)$ (Example 3.1), $x^*(t) = \overline{x(t)}$, $t \in \Omega$.

23.3. $A = C(0, 1)$, $x^*(t) = \overline{x(1-t)}$, $0 \leqslant t \leqslant 1$.

23.4. $A = \mathscr{A}$ (Example 3.9), $x^*(t) = \overline{x(\bar{t})}$, $|t| \leqslant 1$.

23.5. $A = L_1(-\infty, \infty)$ (Example 3.3), $x^*(t) = \overline{x(-t)}$, $-\infty < t < \infty$.

23.6. $A = l_1(-\infty, \infty)$ (Example 3.5), $x^*(n) = \overline{x(-n)}$, $-\infty < n < \infty$ or, after passing to the Gelfand transforms, $(x^*)\hat{}(t) = \overline{x\hat{}(t)}$, $|t| = 1$.

23.7. $A = B(H)$ with H a Hilbert space (Example 3.11), the involution being defined as the operation $T \to T^*$ of taking the adjoint operator.

23.8. DEFINITION. A subalgebra A_0 of a Banach algebra with an involution is called a *self-adjoint subalgebra* provided $A_0^* = A_0$, i.e. if the subalgebra A_0 is closed under involution. A self-adjoint subalgebra will be also referred to as a *-subalgebra. We shall also refer to algebras with an involution as *self-adjoint algebras*, in symbols **-algebras*. *Self-adjoint homomorphisms* of such algebras (or **-homomorphisms*) denote homomorphisms preserving the involution, i.e. such that $h(x^*) = [h(x)]^*$.

In the rest of this section only Banach algebras with an involution will be considered.

Example 23.7 is one of the most important examples of an algebra with an involution. We shall see in the sequel that an important class of algebras with an involution consists precisely of algebras isomorphic to *-subalgebras of this algebra.

We introduce the following terminology by analogy to the theory of operators in Hilbert spaces.

23.9. DEFINITION. An element x of an *-algebra A is called *normal* if $x^*x = xx^*$. Thus an element $x \in A$ is normal if and only if it is contained in a commutative *-subalgebra of A.

23.10. DEFINITION. An element $x \in A$ is called *Hermitian* if $x = x^*$. Every Hermitian element is normal; moreover, every element $x \in A$ can be written as the sum

$$x = a + ib, \tag{23.10.1}$$

where a and b are Hermitian elements of A. (To see this, it suffices to set $a = \frac{1}{2}(x+x^*), b = \frac{1}{2i}(x-x^*)$.) The set of all Hermitian elements of an algebra A will be denoted by H_A.

Similarly a subset $X \subset A$ will be called *Hermitian* (*self-adjoint*) if $X = \{x^*: x \in X\}$.

23.10.a. EXERCISE. Prove that the unit e of an algebra with an involution is a Hermitian element.

23.10.b. EXERCISE. Prove that an element x of an *-algebra A with unit is invertible if and only if the element x^* is invertible and that $(x^{-1})^* = (x^*)^{-1}$. In particular, the inverses of Hermitian elements are Hermitian.

23.10.c. EXERCISE. Prove that the product of Hermitian elements is Hermitian if and only if those elements commute.

23.11. DEFINITION. A Hermitian element $p \in A$ is called a *projection* if it is an idempotent, i.e. if $p^2 = p$.

23.11.a. EXERCISE. Prove that if an algebra A has a unit, then $e-p$ is a projection if and only if p is a projection.

23.12. THEOREM. *If A is an algebra with a continuous involution $x \to x^*$, then there exists in A an equivalent submultiplicative norm satisfying the condition*

$$||x^*|| = ||x|| \qquad (23.12.1)$$

and $||e|| = 1$ provided A has a unit.

PROOF. It is easy to see that if $|||x|||$ is the original norm in A and the involution $x \to x^*$ is continuous, then the norm $||x|| = \max(|||x|||, |||x^*|||)$ enjoys all required properties. The equality $||e|| = 1$ follows from the fact that e is a Hermitian element (Exercise 23.11.a) and $|||e||| = 1$.

Since in the sequel only algebras with continuous involution will be consideredi we shall assume that condition (23.12.1) is fulfilled. The reader can easily verify that the natural norms in Examples 23.2–23.7 satisfy this condition.

23.13. DEFINITION. A commutative algebra with an involution is called a *symmetric algebra* if

$$(x^*)\hat{}(M) = \overline{x\hat{}(M)}$$

for all $M \in \mathfrak{M}(A)$. The involution $x \to x^*$ is then called a *symmetric involution*.

23.13.a. EXERCISE. Prove that Examples 23.2 and 23.6 are symmetric algebras and 23.3, 23.4, 23.5 are non-symmetric.

23.13.b. EXERCISE. Prove, with use of the closed graph theorem, that the involution in semi-simple commutative Banach algebras is continuous.

23.14. THEOREM. *A commutative *-algebra A with unit is a symmetric algebra if and only if for every element $x \in A$ the inverse $(e+x^*x)^{-1}$ exists.*

PROOF. If A is a symmetric algebra, then $(e+x^*x)\hat{}(M) = 1+|x\hat{}(M)|^2 > 0$ so, by Theorem 8.11, the element $e+x^*x$ is invertible in A. Conversely, suppose that this inverse exists for every $x \in A$. We show first that the spectrum of a Hermitian element $h \in A$ is real. In fact, for real α and β we have

$$[h-(\alpha+i\beta)e]\,[h-(\alpha-i\beta)e] = (h-\alpha e)^2+\beta^2 e = \beta^2(e+z^*z),$$

where $z = h-\alpha e/\beta$. If $\beta \neq 0$, then the right-hand element is invertible and so are both factors on the left-hand side, whence it follows that $\alpha+\beta i \notin \sigma(h)$. Hence we may write any element $x \in A$ in the form (23.10.1), $x = a+ib$ with a, b Hermitian. Then we have $x^* = a-ib$, so

$$(x^*)\hat{}(M) = a\hat{}(M)-ib\hat{}(M) = \overline{(a+ib)\hat{}(M)} = \overline{x\hat{}(M)},$$

since $a\hat{}(M)$ and $b\hat{}(M)$ are real numbers.

In view of this theorem we may generalize Definition 23.13 to the case of non-commutative algebras:

23.15. DEFINITION. An *-algebra A with unit is called a *symmetric algebra* if for every $x \in A$ there exists the inverse $(e+x^*x)^{-1} \in A$.

§ 24. B*-ALGEBRAS AND THE GELFAND–NAIMARK THEOREM

An important class of algebras with an involution is that of the so-called **B***-algebras.

24.1. DEFINITION. A Banach algebra A with an involution $x \to x^*$ is called a **B***-*algebra* if its topology may be defined by means of a submultiplicative norm which also satisfies the condition

$$(24.1.1) \qquad ||x^*x|| = ||x||^2$$

for all $x \in A$.

REMARK. Such a norm necessarily also satisfies condition (23.12.1). For we have $||x^2|| = ||x^*x|| \leqslant ||x^*||\,||x||$ whence $||x|| \leqslant ||x^*||$. Analogously $||x^*|| \leqslant ||x||$ and so $||x^*|| = ||x||$.

24.1.a. EXERCISE. Prove that condition (24.1.1) is equivalent to the conditions $||x^*|| = ||x||$ and $||x^*x|| = ||x^*||\,||x||$.

24.1.b. EXERCISE. Prove that if a submultiplicative norm in an *-algebra A satisfies the conditions

$$(24.1.2) \qquad ||x^*|| = ||x|| \quad \text{and} \quad ||x^*x|| \geqslant ||x||^2,$$

then it satisfies also (24.1.1) and A is a **B***-algebra. (This is an extension of the preceding exercise.)

An example of a commutative **B***-algebra is given by 23.2. We shall show that the algebra $B(H)$ (Example 23.7) is a non-commutative **B***-algebra. Indeed, a Hilbert space H is isometrically isomorphic to its dual so that for any $\xi \in H$ we have

$$||\xi|| = \sup_{||\eta|| \leqslant 1} |\langle \xi, \eta \rangle|, \quad \text{where } ||\xi||^2 = \langle \xi, \xi \rangle.$$

Thus for every operator $T \in B(H)$ we have

$$||T|| = \sup_{||\xi|| \leqslant 1} ||T\xi|| = \sup_{||\xi||, ||\eta|| \leqslant 1} |\langle T\xi, \eta \rangle|.$$

Hence

$$||T^*T|| = \sup_{||\xi||, ||\eta|| \leqslant 1} |\langle T^*T\xi, \eta \rangle| = \sup_{||\xi||, ||\eta|| \leqslant 1} |\langle T\xi, T\eta \rangle|$$
$$\geqslant \sup_{||\xi|| \leqslant 1} |\langle T\xi, T\xi \rangle| = ||T||^2,$$

and since $||T|| = ||T^*||$, relations (24.1.2) hold and $B(H)$ is a **B***-algebra.

We are going to show that the two given examples are in a sense universal: every commutative **B***-algebra with unit is an algebra of type $C(\Omega)$ and every (possibly non-commutative) **B***-algebra is isometrically isomorphic to a self-adjoint subalgebra of the algebra $B(H)$. The involution is in the first case the operation of taking the conjugate function and in the second case the operation of taking the adjoint operator. Now we shall prove the first of these statements.

24.2. The Gelfand–Naimark Theorem. *If A is a commutative* **B***-*algebra with unit, then the Gelfand homomorphism $x \to x^\wedge$ is an isometric isomorphism of A onto the algebra $C(\mathfrak{M})$ with $\mathfrak{M} = \mathfrak{M}(A)$; the images of elements x and x^* are conjugate functions. In particular, A is a symmetric and semi-simple algebra.*

Proof. We first prove that the mapping $x \to x^\wedge$ is an isometry. Let $x \in A$; we have

$$||x^2||^2 = ||x^{2*}x^2|| = ||(x^*x)^*(x^*x)|| = ||x^*x||^2 = ||x||^4$$

so that $||x^2|| = ||x||^2$ and, by induction, $||x^{2^n}|| = ||x||^{2^n}$. Thus

$$||x_s|| = \lim \sqrt[n]{||x^n||} = \lim \sqrt[2^n]{||x^{2^n}||} = ||x||,$$

so the injection $A \to A^\wedge \subset C(\mathfrak{M})$ is an isometry and A is semi-simple.

Now we prove that for any $M \in \mathfrak{M}$ we have $(x^*)^\wedge(M) = \overline{x^\wedge(M)}$. Fix an $x \in A$ and $M \in \mathfrak{M}$. Let $x^\wedge(M) = \alpha+\beta i$ and $(x^*)^\wedge(M) = \gamma+\delta i$. We are to show that $\alpha = \gamma$ and $\beta = -\delta$. We start with the last equality. Assume that $\beta+\delta \neq 0$ and put $y = (\beta+\delta)^{-1}[x+x^*-(\alpha+\gamma)e]$. Clearly $y^* = y$ and $y^\wedge(M) = i$. For any real number r we have

$$(y+ire)^\wedge(M) = i(1+r),$$

whence

$$|1+r|^2 \leqslant ||y+ire||_s^2 = ||y+ire||^2 = ||(y+ire)^*(y+ire)||$$
$$= ||(y-ire)\,(y+ire)|| = ||y^2+r^2e|| \leqslant ||y^2||+r^2$$

and the contradiction is obtained by putting $r = ||y^2||$. Hence $\beta+\delta = 0$ and $x^\wedge(M) = \alpha+\beta i$, $(x^*)^\wedge(M) = \gamma-\beta i$. Turning to the element ix we have $(ix)^\wedge(M) = -\beta+\alpha i$, $[(ix)^*]^\wedge(M) = -i(x^*)^\wedge(M) = \beta-\gamma i$ and repeating the argument we get $\alpha = \gamma$. Thus the involution in A is symmetric. Now, together with any function $x^\wedge \in A^\wedge$ the conjugate function $\overline{x^\wedge}$ is also in $A^\wedge$, the functions $x^\wedge$ separate the points of $\mathfrak{M}$ and the norm in A is identical with the spectral norm and is complete. Consequently, by the Weierstrass–Stone Theorem, $A = C(\mathfrak{M})$.

24.3. Corollary. *Every commutative* **B***-*algebra without unit is isometrically isomorphic to an algebra $C_0(\mathfrak{M})$ of all continuous functions on $\mathfrak{M} = \mathfrak{M}(A)$ "vanishing at infinity" and the involution is the operation of taking the conjugate function.*

24.4. Corollary. *The spectrum of a Hermitian element of a* **B***-*algebra is real.*

24.5. Coorllary. *For every Hermitian element x of a* **B***-*algebra we have*

$$||x||_s = ||x||. \tag{24.5.1}$$

24.6. Theorem. *If A is a* **B***-*algebra with a unit e and if A_0 is a self-adjoint subalgebra of A containing e, then for any $x \in A_0$ we have $\sigma_{A_0}(x) = \sigma_A(x)$.*

Proof. We first show that every element of A_0 invertible in A is also invertible in A_0. Let $z \in A_0$, $z^{-1} \in A$. Then there exists the inverse $(z^*)^{-1} \in A$ (see Exercise

23.10.b) and z^*z is an Hermitian element of A_0, invertible in A. In virtue of Corollary 23.4 its spectrum is real; moreover, this spectrum does not contain 0. We thus have, on account of Corollary 17.6, $\sigma_A(z^*z) = \sigma_{A_0}(z^*z)$. It hence follows that the element z^*z, and thus also z, is invertible (¹) in A_0.

Since the spectrum $\sigma_{A_0}(x)$ (or $\sigma_A(x)$) depends only on the invertibility in A_0 (in A) of the elements $x - \lambda e$ and since such elements are or are not invertible simultaneously in both algebras then the spectra $\sigma_A(x)$ and $\sigma_{A_0}(x)$ are identical.

This theorem reduces problems concerning the spectra of normal elements x to problems involving their spectra in commutative subalgebras generated by x and x^*.

24.7. Remark. The above definition of a **B***-algebra is sometimes not quite manageable in applications. For if a norm satisfying (24.1.1) is replaced by another equivalent submultiplicative norm, then this new norm satisfies the condition

(24.7.1) $$\|x^*\| \, \|x\| \leqslant C\|x^*x\|$$

instead of (24.1.1).

We shall show that if the norm of a commutative *-algebra A satisfies condition (24.7.1), then A is a **B***-algebra (this is also true in the non-commutative case; see Remark 26.13).

24.8. Theorem. *If A is a commutative *-algebra with unit and its norm satisfies condition* (24.7.1), *then A is a* **B***-*algebra.*

Proof. We may assume that the norm of A satisfies also (23.12.1). It suffices to prove that the mapping $x \to x^\wedge$ is an isomorphism of A onto $C(\mathfrak{M})$, where $\mathfrak{M} = \mathfrak{M}(A)$ and that $(x^*)^\wedge(M) = \overline{x^\wedge(M)}$ for all $x \in A$ and $M \in \mathfrak{M}$. Let $x \in A$. Applying inequality (24.7.1) to the powers x^n we obtain $\|(x^*)^n\| \, \|x^n\| \leqslant C\|(x^*x)^n\|$. Taking n-th roots on both sides and passing to the limit $n \to \infty$ we obtain $\|x^*\|_s \|x\| \leqslant \|x^*x\|_s$. Also, $\|x^n\| = \|(x^*)^n\|$ whence $\|x\|_s = \|x^*\|_s$ and consequently $\|x^*x\|_s = \|x\|_s^2$. In view of this last equality we may repeat the argument occurring in the proof of the Gelfand–Naimark Theorem (Theorem 24.2) which shows that $(x^*)^\wedge(M) = \overline{x^\wedge(M)}$ for all $x \in A$ and $M \in \mathfrak{M}$. By the Weierstrass–Stone Theorem, $A^\wedge$ is dense in $C(\mathfrak{M})$ in the norm $\|x\|_s$. The proof will be complete if we show that the norms $\|x\|$ and $\|x\|_s$ are equivalent. If $z \in A$ is a Hermitian element, then we get from (24.7.1) $\|z\|^2 \leqslant C\|z^2\|$ whence, by induction, $\|z\|^{2^n} \leqslant C^{2^n-1}\|z^{2^n}\|$ or $\|z\| \leqslant C^{(2^n-1)/2^n}\|z^{2^n}\|^{2^{-n}}$. Passing to the limit as $n \to \infty$ we get $\|z\| \leqslant C\|z\|_s$. If now x is an arbitrary element of A, then

$$\|x\|^2 = \|x^*\| \, \|x\| \leqslant C\|x^*x\| \leqslant C^2\|x^*x\|_s = C^2\|x\|_s^2$$

(¹) If $a, b \in A$ and ab, ba are invertible in A, then so are a and b. For then there exist $p, q \in A$ such that $abp = e$ and $qba = e$ so that a is both left- and right-invertible, hence is invertible in A. But the invertibility of ab alone does not imply that of a (see the elements occurring in Exercise 17.7.a).

and thus $||x|| \leqslant C||x||_s$ which together with $||x||_s \leqslant ||x||$ gives the desired equi valence of norms.

24.9. Remark. The above theorem is valid also for algebras without unit i a formulation analogous to that of Corollary 24.3. It is also true for non-com mutative **B***-algebras, cf. Yood [2].

§ 25. THE THEORY OF REPRESENTATIONS OF ALGEBRAS WITH AN INVOLUTION

We now pass to the theory of representations of algebras with an involutior It provides a tool for the proof (in the next section) of the theorem about th isomorphism between **B***-algebras and closed self-adjoint subalgebras of $B(H)$ This theory will be continued in subsequent sections.

25.1. Definition. Let H be a complex Hilbert space and let $B(H)$ denote th algebra of all endomorphisms of this space. Every *-homomorphism of an *-algebr A into the algebra $B(H)$ will be called a *representation of the algebra A* in th space H. Thus a representation is just a transformation $x \to T_x$ of A into $B(H$ satisfying the following conditions:

(i) $T_{\alpha x+\beta y} = \alpha T_x + \beta T_y$,

(ii) $T_{xy} = T_x T_y$,

(iii) $T_{x^*} = T_x^*$,

(iv) $T_e = I$ whenever A has a unit; I is here the identity operator of H.

25.2. Definition. A representation $x \to T_x$ of an algebra A in a space H called a *cyclic representation* if there exists an element $\xi_0 \in H$, called the *cycl vector of this representation*, such that the set $\{T_x \xi_0 : x \in A\}$ is dense in H. Th set is evidently a linear subset of H.

25.3. Definition. Two representations $x \to T_x$ and $x \to T'_x$ of an algebra A i spaces H resp. H' are said to be *equivalent* if there exists a unitary operator of H onto H' such that

$$T'_x S\xi = T_x \xi$$

for all $x \in A$ and $\xi \in H$.

25.4. Definition. We say that a subspace $H_1 \subset H$ is an *invariant subspace* a representation $x \to T_x$ if $T_x H_1 \subset H_1$ for every $x \in A$. In this case the restri tion $x \to T_x|_{H_1}$ also is a representation of A in the space H_1. A representatic $x \to T_x$ is called *irreducible* if there exists no proper (i.e. distinct from $\{0\}$ ar H) closed invariant subspace.

25.4.a. Exercise. Prove that a representation equivalent to an irreducible representati is itself irreducible.

25.4.b. Exercise. Prove that if H_1 is an invariant subspace of a representation $x \to T$ then its orthogonal complement $H_2 = H_1^\perp$ is also invariant.

25.4.c. Exercise. Prove, applying the preceding exercise, that a space H of a representation $x \to T_x$ admits a decomposition into a direct sum of a family of invariant subspaces H_α such that the restricted representations $x \to T_x|_{H_\alpha}$ are cyclic.

Hint. Apply transfinite induction or the Kuratowski–Zorn lemma.

25.5. Theorem. *A representation $x \to T_x$ of an *-algebra A in a space H is irreducible if and only if for every operator $S \in B(H)$ the condition $ST_x = T_xS$ (for all $x \in A$) implies $S = \lambda I$, $\lambda \in \mathbf{C}$.*

Proof. (a) Assume that the representation $x \to T_x$ is irreducible and that an operator S commutes with all T_x's. Assume further that S is self-adjoint and consider the subalgebra $A_S = [S, I] \subset B(H)$. This is a commutative **B***-algebra, so, by Theorem 24.2, A_S is isomorphic to $C(\sigma(S))$, as $\mathfrak{M}(A_S) = \sigma(S)$ (Theorem 10.11). We shall show that the spectrum $\sigma(S)$ is reduced to a single point and thus, by the semi-simplicity of A_S, $S = \lambda I$.

Suppose that $\sigma(S)$ contains at least two points; then the algebra $C(\sigma(S))$ has divisors of zero and so there exist operators $U, V \in A_S$, $U \neq 0 \neq V$, such that $UV = VU = 0$. Let H_1 be the closure of $U(H)$, a subspace of H. Consider an element $\xi = V^*\varphi \neq 0$ (which certainly exists since $V \neq 0$). Let $\eta \in U(H)$, $\eta = U\psi$, $\psi \in H$. We have $\langle \xi, \eta \rangle = \langle V^*\varphi, U\psi \rangle = \langle \varphi, VU\psi \rangle = 0$. Thus $\xi \perp U(H)$, hence also $\xi \perp H_1$ and $H_1 \neq H$. Now, all operators in A_S commute with T_x's and thus we have $T_x U(H) = UT_x(H) \subset U(H)$ so that $T_x(H_1) \subset H_1$ for all $x \in A$, a contradiction to the irreducibility of the representation.

If S is not Hermitian, then the above argument may be applied to the Hermitian operators

$$\frac{1}{2}(S+S^*), \quad \frac{1}{2i}(S-S^*)$$

and the condition $S = \lambda I$ also follows. It should be noted that the operator S^* also commutes with all T_x's, since

$$S^*T_x = (T_{x^*}S)^* = (ST_{x^*})^* = T_xS^*.$$

(b) Now assume that the representation $x \to T_x$ is not irreducible and let H_1 be a proper invariant subspace of H. Let P_1, resp. P_2 denote the projections of H onto H_1, resp. $H_2 = H_1^\perp$, the orthogonal complement of H_1. For every $\xi \in H$ we have $T_xP_1\xi \in H_1$, so $P_1T_xP_1\xi = T_xP_1\xi$ and

$$P_1T_xP_1 = T_xP_1 \tag{25.5.1}$$

for all $x \in A$. Further, $P_1+P_2 = I$, so

$$P_1T_x = P_1T_x(P_1+P_2) = P_1T_xP_1+P_1T_xP_2. \tag{25.5.2}$$

H_2 is also an invariant subspace for the representation $x \to T_x$ (Exercise 25.4.b), so we have, by (25.5.1), $P_2T_xP_2 = T_xP_2$ for all $x \in A$. Since $P_1P_2 = P_1(I-P_2) = P_1-P_1 = 0$, we may rewrite (25.5.2) as

$$P_1T_x = P_1T_xP_1+P_1P_2T_xP_2 = P_1T_xP_1. \tag{25.5.3}$$

Comparing relations (25.5.1) and (25.5.3) we see that $P_1 T_x = T_x P_1$ holds for every $x \in A$. Thus the operators T_x commute with an operator which certainly is not a scalar multiple of the identity and the proof is complete.

We now introduce the important notion of a positive functional.

25.6. DEFINITION. An additive and homogeneous functional f defined on an *-algebra A is a *positive functional* if

$$f(x^*x) \geqslant 0 \tag{25.6.1}$$

for every element $x \in A$.

25.6.a. EXERCISE. Prove that if f is a positive functional defined on an algebra A, then the following inequality (an analogue of the Schwarz inequality)

$$|f(y^*x)|^2 \leqslant f(y^*y)f(x^*x) \tag{25.6.2}$$

holds for all $x, y \in A$.

25.7. DEFINITION. A linear functional f defined on an *-algebra A is called a *real functional* if $f(x^*) = \overline{f(x)}$ for every $x \in A$.

25.7.a. EXERCISE. Prove that every element f of the dual space A' may be uniquely written in the form

$$f = f_1 + if_2,$$

where f_1 and f_2 are real functionals in A.

Observe that in Definition 25.6 we did not assume the continuity of f, but it follows from the following theorem.

25.8. THEOREM. *Every positive functional f is real and continuous; moreover,*

$$|f(x)| \leqslant f(e)\,||x|| \tag{25.8.1}$$

for every $x \in A$.

PROOF. Assume first that $x = x^*$ and $||x|| < 1$. The spectrum of the element $e-x$ is thus contained in the disc $\{\xi \in \boldsymbol{C}\colon |1-\xi| < 1\}$ in which a holomorphic branch of the square root is defined. In virtue of Theorem 16.4 there exists an element $y \in A$ such that $y^2 = e-x$ and $y^* = y$. (The element y may be constructed by means of the power-series expansion of the function $\sqrt{1-\xi}$. Its coefficients are real, so applying this series to x we obtain a Hermitian element). Thus $y^*y = e-x$, so $f(e)-f(x) = f(e-x) = f(y^*y) \geqslant 0$. Similarly $f(e)+f(x) = f(e)-f(-x) \geqslant 0$. It follows that $f(x)$ and $f(e)$ are real numbers and $|f(x)| \leqslant f(e)$. Now, if y is any self-adjoint element and if $\delta > 1$, then $||y/\delta||y||\,|| < 1$ and $|f(y/\delta||y||)| \leqslant f(e)$ whence $|f(y)| \leqslant \delta f(e)\,||y||$ and since this holds for any $\delta > 1$, so $|f(y)| \leqslant f(e)\,||y||$. In particular,

$$f(x^*x) \leqslant f(e)\,||x^*x|| \leqslant f(e)\,||x||^2 \tag{25.8.2}$$

for every $x \in A$.

On the other hand, if we put $y = e$ in (25.6.2), we obtain $|f(x)|^2 \leqslant f(e)f(x^*x)$. This inequality together with (25.8.2) gives $|f(x)|^2 \leqslant [f(e)]^2\|x\|^2$, which is just (25.8.1).

It remains to verify that f is a real functional. As shown above, $f(y)$ is a real number for Hermitian y. Thus, writing

$$f(x) = f\left(\frac{x+x^*}{2}\right) + if\left(\frac{x-x^*}{2i}\right), \quad f(x^*) = f\left(\frac{x+x^*}{2}\right) - if\left(\frac{x-x^*}{2i}\right),$$

we get $f(x^*) = \overline{f(x)}$, since the elements $\dfrac{x+x^*}{2}$ and $\dfrac{x-x^*}{2i}$ are Hermitian.

25.8.a. EXERCISE. Let $x \to T_x$ be a representation of an *-algebra A in a Hilbert space H. Prove that for a fixed $\xi \in H$ the functional

(25.8.3) $$f(x) = \langle T_x\xi, \xi\rangle$$

is positive.

We now give a stronger version of the result of Theorem 25.8 stating that positive functionals are continuous in the presence of a unit element.

25.9. THEOREM. *Let A be an *-algebra with a (bounded) left approximate identity and a continuous involution. Then every positive functional on A is continuous.*

PROOF. Let f be a positive functional on A and put $A_1 = A \oplus \{\lambda e\}$. If we set $(x+\lambda e)^* = x^* + \bar{\lambda}e$, A_1 becomes also an *-algebra. We now define a positive functional F on A_1 setting

(25.9.1) $$F(x) = f(a^*xa),$$

where a is a fixed element in A. By Theorem 25.8 it is a continuous functional on A_1. If we use the formula

$$4axb = \sum_{k=0}^{3} i^k(a^* + i^{-k}b)^*x(a^* + i^{-k}b)$$

we obtain

(25.9.2) $$4f(axb) = \sum_{k=0}^{3} i^k f[(a^* + i^{-k}b)^*x(a^* + i^{-k}b)]$$

and so $x \to f(axb)$ is a continuous functional defined for $x \in A$.

Let now $x_n \to 0$ in A. By Theorem 6.4 we can write $x_n = ay_n$, where $a \in A$ and $y_n \to 0$. By the right version of Theorem 6.4 (cf. Exercise 6.4.a) we can write $y_n = z_n b$, where $z_n \to 0$ and $b \in A$. This is justified, since, as can be easily verified, (δ_α^*) is a right approximate identity for A if δ_α is a left one. We have now $x_n = az_n b$ and $z_n \to 0$, so $f(x_n) = f(az_n b) \to 0$ since the functional $x \to f(axb)$ is continuous in A, but this means that the functional f is itself continuous (1).

(1) As shown recently in Murphy [1], Theorem 25.9 is also true without the assumption of continuity of the involution.

25.9.a. EXERCISE. Consider the algebra l_p of Exercise 6.5.a. It is an *-algebra with a continuous involution if for $x = (x_\alpha) \in l_p$ we set $x^* = (\bar{x}_\alpha)$. Prove that there are discontinuous positive functionals in l_p, which shows that the condition of boundedness imposed on an approximative identity is essential in the proof of Theorem 25.9 (cf. Exercise 6.5.a).

25.10. THEOREM. *Every representation $x \to T_x$ of an *-algebra A is continuous; moreover,*

$$||T_x|| \leqslant ||x|| \tag{25.10.1}$$

for every $x \in A$.

PROOF. Applying inequality (25.8.2) to functional (25.8.3) we obtain

$$\langle T_{x^*x}\xi, \xi\rangle \leqslant \langle \xi, \xi\rangle ||x||^2,$$

hence $||T_x\xi||^2 \leqslant ||\xi||^2||x||^2$ and (25.10.1) follows.

Now we shall be concerned with connections between the positive functionals and cyclic representations of *-algebras.

25.11. THEOREM. *Let A be an algebra with an involution; let $x \to T_x$ be a cyclic representation of A in a space H and ξ_0 its cyclic vector. Then the functional*

$$f(x) = \langle T_x\xi_0, \xi_0\rangle \tag{25.11.1}$$

is positive and determines the representation uniquely up to an equivalence.

Conversely, *to every positive functional f defined on A there corresponds a cyclic representation $x \to T_x$ of A in some Hilbert space H where the connection between the functional and the representation is given by formula* (25.11.1).

PROOF. The functional f given by (25.11.1) is positive (see Exercise 25.6.a). It remains to show that if $f(x) = \langle T_x\xi_0, \xi_0\rangle = \langle T'_x\xi'_0, \xi'_0\rangle'$, where $x \to T_x$, $x \to T'_x$ are cyclic representations of A in spaces H, H' with scalar products $\langle,\rangle$, $\langle,\rangle'$ and with cyclic vectors ξ_0, ξ'_0 respectively, then the two representations $x \to T_x$, $x \to T'_x$ are equivalent. Let $\xi_x = T_x\xi_0$ and write $\xi'_x = S\xi_x = T'_x\xi'_0$. We shall show that S is a unitary operator which maps the dense linear subset $Z = \{\xi_x \in H: x \in A\} \subset H$ into H'. Indeed,

$$\begin{aligned}\langle S\xi_x, S\xi_y\rangle' &= \langle T'_x\xi'_0, T'_y\xi'_0\rangle' = \langle T'_{y^*x}\xi'_0, \xi'_0\rangle' = f(y^*x)\\ &= \langle T_{y^*x}\xi_0, \xi_0\rangle = \langle T_x\xi_0, T_y\xi_0\rangle = \langle \xi_x, \xi_y\rangle.\end{aligned}$$

The set Z is dense in H since the representation is cyclic, thus the operator S can be extended by continuity onto the whole of H. The space H is then carried onto entire H', since the set $SZ = \{\xi'_x \in H': x \in A\}$ is dense in H' and S is an isometry. For any $\xi_x \in Z$ we have further

$$ST_y\xi_x = ST_{yx}\xi_0 = S\xi_{yx} = \xi'_{yx} = T'_{yx}\xi'_0 = T'_yT'_x\xi'_0 = T'_y\xi'_x = T'_yS\xi_x,$$

whence $ST_y = T'_yS$ since Z is dense in H. We have shown that the two representations are equivalent and the first assertion is thus proved.

Now suppose that f is a positive functional defined in A. We have to construct a Hilbert space H and a cyclic representation $x \to T_x$ such that (25.11.1) holds. For this purpose we define a bilinear form $\langle x, y \rangle = f(y^*x)$ in A. It is easily seen that the set $I = \{x \in A: \langle x, x \rangle = 0\}$ is a left ideal in A. The space A/I is a pre-Hilbert space with the scalar product $\langle \xi, \eta \rangle = \langle x, y \rangle$, where $x \in \xi$, $y \in \eta$, since the value $\langle \xi, \eta \rangle$ does not depend on the choice of x and y in the cosets ξ, η. Let H denote the completion of A/I in the norm $\|\xi\| = \langle \xi, \xi \rangle^{1/2}$. We define a representation $x \to T_x$ by $T_x\xi = \eta$, where ξ is an element of A/I and η is the coset containing the set $x \cdot \xi$. (The existence of such a coset follows from the fact that I is an ideal.) We shall show that $\|T_x\| \leqslant \|x\|$. In fact, $\langle T_x\eta, T_x\eta \rangle = f(y^*x^*xy)$, where $y \in \eta \in A/I$. For a fixed y the functional $f_1(x) = f(y^*xy)$ is a positive functional in A. Hence, by formula (25.8.2), $f_1(x^*x) \leqslant f_1(e)\|x\|^2$, so that $f(y^*x^*xy) \leqslant f(y^*y)\|x\|^2$, i.e. $\langle T_x\eta, T_x\eta \rangle \leqslant \langle \eta, \eta \rangle \|x\|^2$ and $\|T_x\| \leqslant \|x\|$. The operators T_x, being bounded, can be extended onto the whole of H and the resulting extensions define a representation of A in H, as can be easily verified. To see that this representation is cyclic it suffices to check that the coset $\xi_0 \in A/I$ containing the unit of A is a cyclic vector. Finally, $\langle T_x\xi_0, \xi_0 \rangle = f(e^*x) = f(x)$.

25.11.a. EXERCISE. Complete the above proof by proving in detail all those facts whose proofs have been omitted.

We shall conclude this section with the important notion of the direct sum of a family of representations.

25.12. DEFINITION. Let $x \to T_x^\alpha$ be a family of representations of an *-algebra A in spaces H_α with scalar products $\langle , \rangle_\alpha$ respectively, $\alpha \in \mathfrak{A}$, an index set. Let H denote the Hilbert space whose elements are systems $\xi = (\xi_\alpha)_{\alpha\in\mathfrak{A}}$, $\xi_\alpha \in H_\alpha$ such that

$$\|\xi\|^2 = \sum_{\alpha\in\mathfrak{A}} \|\xi_\alpha\|_\alpha^2 < \infty .$$

The representation $x \to T_x$ of A in H given by

$$T_x\xi = (T_x^\alpha \xi_\alpha)_{\alpha\in\mathfrak{A}}$$

is called the *direct sum of the family of representations* $x \to T_x^\alpha$, $\alpha \in \mathfrak{A}$. The operators T_x are bounded in H since, by (25.10.1),

$$\|T_x\xi\|^2 = \sum_\alpha \|T_x^\alpha \xi_\alpha\|_\alpha^2 \leqslant \|x\|^2 \sum_\alpha \|\xi_\alpha\|_\alpha^2 = \|x\|\,\|\xi\|^2.$$

If S_{α_0} denotes the isometric injection of H_{α_0} into H given by $\xi_{\alpha_0} \to (\eta_\alpha)_{\alpha\in\mathfrak{A}}$, where $\eta_{\alpha_0} = \xi_{\alpha_0}$, $\eta_\alpha = 0$ for $\alpha \neq \alpha_0$, then $T_x S_\alpha \xi_\alpha = T_x^\alpha \xi_\alpha$ and the image $S_\alpha(H^\alpha)$ is an invariant subspace of the representation $x \to T_x$.

The notion of the direct sum of a family of representations will be used in the next section in the proof of the announced theorem on the isomorphism between any **B***-algebra and a self-adjoint subalgebra of the algebra $B(H)$.

§ 26. THE REPRESENTATION THEOREM FOR B*-ALGEBRAS

We shall prove now that *every* **B***-algebra is isometrically isomorphic to some *-subalgebra of the algebra* $B(H)$. Self-adjoint (norm-closed) subalgebras of $B(H)$ are generally called in the literature **C***-*algebras*, thus the theorem may be reformulated as follows: *every* **B***-*algebra is a* **C***-*algebra*. Of course, every **C***-algebra is automatically a **B***-algebra so that the two classes coincide. The expression "**B***-algebra" usually refers to an "abstract" (axiomatically defined) algebra whereas the name "**C***-algebra" is used with regard to a specific *-subalgebra of $B(H)$.

The following theorem reduces the problem to the case of algebras with unit.

26.1. THEOREM. *Let A be a* **B***-*algebra. If A has no unit, then it can be isometrically *-isomorphically embedded in a* **B***-*algebra $\tilde{A}$ with unit.*

PROOF. Let $||\cdot||$ be a submultiplicative norm in A satisfying condition (24.1.1). To every $x \in A$ we assign the operator l_x: $l_x y = xy$, mapping A into itself. We have

$$||l_x|| = \sup_{y \neq 0} \frac{||xy||}{||y||} \leqslant ||x||.$$

On the other hand

$$||x||^2 = ||x^*x|| = ||(x^*x)^*|| = ||xx^*|| = ||l_x x^*|| \leqslant ||l_x|| \, ||x^*|| = ||l_x|| \, ||x||,$$

so that $||x|| \leqslant ||l_x||$, whence $||x|| = ||l_x||$ and the injection $x \to l_x$ of A into $B(A)$ is an isometric isomorphism. Let I denote the identity operator of $B(A)$ and write

$$\tilde{A} = \{l_x + \alpha I \in B(A) : x \in A, \alpha \in C\}.$$

$\tilde{A}$ is a closed subalgebra of $B(A)$ with unit and contains A isomorphically and isometrically. The involution in $\tilde{A}$ is defined by

$$T^* = l_{x^*} + \bar{\alpha} I$$

if $T = l_x + \alpha I$. We shall show that $\tilde{A}$ is a **B***-algebra under this involution. We estimate

$$||Ty||^2 = ||xy + \alpha y||^2 = ||(xy+\alpha y)^*(xy+\alpha y)|| = ||y^*(T^*Ty)||$$
$$\leqslant ||y^*|| \, ||T^*Ty|| \leqslant ||T^*T|| \, ||y||^2.$$

Hence it follows

$$||T||^2 = (\sup_{||y|| \leqslant 1} ||Ty||)^2 \leqslant ||T^*T|| \leqslant ||T^*|| \, ||T|| \tag{26.1.1}$$

and $||T|| \leqslant ||T^*||$ whence, by symmetry, $||T^*|| \leqslant ||T||$. Thus $||T|| = ||T^*||$ which together with (26.1.1) gives the desired relation $||T^*T|| = ||T||^2$. Finally, the assertion follows from the observation that our injection of A into $\tilde{A}$ is also an *-isomorphism.

26.2. DEFINITION. Let X be a vector space over the field of the reals and let K be a cone in X, i.e. a subset of X such that the conditions $x, y \in K$, $\alpha, \beta > 0$ yield $\alpha x+\beta y \in K$. We do not require that K should be a proper cone, i.e. one which contains no subspace of X. An example of a cone is any set $\{x \in X: f(x) \geqslant 0\}$ with f a real linear functional defined in X. It can be easily seen that the relation $x \prec_K y$ defined by the condition $y-x \in K$ is an ordering in X with the property that if $x \prec_K u$ and $y \prec_K v$, then $\alpha x+\beta y \prec_K \alpha u+\beta v$ for $\alpha, \beta > 0$.

A real linear functional f defined in X is said to be *positive with respect to the cone* K if the relation $x \prec_K y$ implies $f(x) \leqslant f(y)$ or, which is the same, if $f(x) \geqslant 0$ for $x \in K$.

In the sequel we shall need the following lemma on the extension of functionals positive with respect to a cone.

26.3. LEMMA. *Let Y be a vector subspace of a real vector space X and let K be a cone in X. Suppose that f is a functional positive with respect to the cone $K \cap Y$, defined in Y, and that $(z+K) \cap Y \neq \emptyset$ for any $z \in X$. Then there exists a functional F defined in X, positive with respect to K and such that $F(x) = f(x)$ for $x \in Y$.*

PROOF. Let E denote the family of all extensions $\alpha = (X_\alpha, F_\alpha)$ of f satisfying the conditions $Y \subset X_\alpha \subset X$, $F_\alpha(x) = f(x)$ for $x \in Y$ and $F_\alpha(x) \geqslant 0$ for $x \in K \cap X_\alpha$. We shall write $\alpha \prec \beta$ whenever $X_\alpha \subset X_\beta$ and $F_\beta(x) = F_\alpha(x)$ for $x \in X_\alpha$. The relation $\prec$ is an ordering in E and satisfies the conditions of the Kuratowski–Zorn lemma, as can be easily seen. Thus there exist in E elements maximal with respect to $\prec$. Let (X_0, F_0) be such a maximal element. We shall show that $X_0 = X$. Indeed, suppose that there exists an element $z \in X \setminus X_0$ and let X_1 denote the subspace of X spanned by X_0 and z. We define a functional F_1 on X_1 by

$$F_1(x) = \begin{cases} F_0(x) & \text{for} \quad x \in X_0, \\ t & \text{for} \quad x = z, \end{cases}$$

where t is any real number from the interval $\lambda \leqslant t \leqslant \mu$, where

$$\mu = \inf_{\substack{y-z \in K \\ y \in X_0}} F_0(y), \qquad \lambda = \sup_{\substack{z-y \in K \\ y \in X_0}} F_0(y).$$

We have $\mu < \infty$ and $\lambda > -\infty$ since y ranges over sets which, by assumption, are non-void.

This definition will be shown to be correct if we prove that $\lambda \leqslant \mu$. But if we had $\lambda > \mu$, then there would exist elements $y_1, y_2 \in X_0$ such that $F_0(y_1) > F_0(y_2)$ and $z-y_1, y_2-z \in K$. Adding the last two elements we then obtain $y_2-y_1 \in K$ which is impossible since $F_0(y_2-y_1) = F_0(y_2)-F_0(y_1) < 0$. So the functional F_1 is defined and $(X_1, F_1) \in E$, as is easy to see. Thus if we had $X_0 \neq X$, then there would exist in E a proper extension of (X_0, F_0), contrary to the maximality of the latter. Consequently $X = X_0$ and F_0 is the desired extension.

We shall apply Lemma 26.3 to the proof of the theorem on the extension of positive functionals defined on *-subalgebras of a **B***-algebra. The proof will be proceded by several lemmas in which A denotes a **B***-algebra with unit.

26.4. LEMMA. *If x and y are Hermitian elements of A and if $\sigma(x)$, $\sigma(y) \geqslant 0$, then $\sigma(x+y) \geqslant 0$* (here the notation $\sigma(z) \geqslant 0$ means that all numbers in $\sigma(z)$ are real non-negative).

PROOF. In view of (24.5.1) we have $||x|| = ||x||_s$ for Hermitian elements: $||x||_s$ is thus a subadditive norm in the set H_A of all Hermitian elements of A. Fix an $\varepsilon > 0$ and choose a positive λ so that $||x'||_s, ||y'||_s < 1$, where $x' = \lambda(x+\varepsilon e)$ and $y' = \lambda(y+\varepsilon e)$. Then $0 < \sigma(e-x') < 1$ and $0 < \sigma(e-y') < 1$, so $||e-x'||_s < 1$ and $||e-y'||_s < 1$. Thus, in view of the subadditivity of $||\cdot||_s$ on H_A, we have $||e-\frac{1}{2}(x'+y')||_s < 1$. It follows that $\sigma(x'+y') > 0$, the spectrum being real because of Corollary 24.4. Thus $\sigma(x+y) > -2\varepsilon$ and since ε is arbitrary positive so $\sigma(x+y) \geqslant 0$.

26.5. LEMMA. *If $x \in A$ and $\sigma(x^*x) \leqslant 0$, then $x = 0$.*

PROOF. Write $x = h+ik$, according to (23.10.1), where h and k are Hermitian elements of A. We have $x^* = h-ik$ whence $x^*x+xx^* = 2(h^2+k^2)$. Since, because of Theorem 24.2, $\sigma(h^2) \geqslant 0$ for every Hermitian element h, then, by the preceding lemma, $\sigma(x^*x+xx^*) \geqslant 0$. On the other hand we have, by Theorem 17.7, $\sigma(xx^*) \leqslant 0$. Thus $\sigma(-x^*x) \geqslant 0$ and $\sigma(-xx^*) \geqslant 0$. Applying once more Lemma 26.4 to the Hermitian elements x^*x+xx^* and $-xx^*$ we get $\sigma(x^*x) \geqslant 0$ and similarly $\sigma(xx^*) \geqslant 0$. Thus $\sigma(x^*x) = \{0\} = \sigma(xx^*)$ whence $||x||^2 = ||x^*x|| = ||x^*x||_s = 0$.

26.6. LEMMA. *For any $x \in A$ we have $\sigma(x^*x) \geqslant 0$.*

PROOF. Consider the subalgebra $A_0 \subset A$ generated by the Hermitian element x^*x. It is a commutative **B***-algebra with unit, so $A_0 = C(\sigma(x^*x))$, by Theorem 24.2. Thus there exist Hermitian elements h and k with non-negative spectra, $h, k \in A_0$, and such that $x^*x = h-k$ and $hk = 0 = kh$ (it suffices to put $h\hat{}(M) = \max\{(x^*x)\hat{}(M), 0\}$ and $k\hat{}(M) = -\min\{(x^*x)\hat{}(M), 0\}$). Write $y = xk$. We have $y^*y = kx^*xk = k(h-k)k = -k^3$, so $\sigma(y^*y) \leqslant 0$ and, by Lemma 26.5, $y = 0$, or $xk = 0$. It hence follows that $(x^*x)^2 = (x^*x)(h-k) = x^*xh$, so that $((x^*x)\hat{}(M))^2 = (x^*x)\hat{}(M)h\hat{}(M) \geqslant 0$ for all $M \in \mathfrak{M}(A_0)$. Thus $(x^*x)\hat{}(M) = 0$ whenever $h\hat{}(M) = 0$ and so $(x^*x)\hat{}(M) \geqslant 0$ for all $M \in \mathfrak{M}(A_0)$. Consequently $\sigma_{A_0}(x^*x) \geqslant 0$ and, according to Theorem 24.6, $\sigma_A(x^*x) = \sigma_{A_0}(x^*x) \geqslant 0$.

26.7. COROLLARY. *Every **B***-algebra with unit is symmetric* (see Definition 23.15).

26.8. LEMMA. *If x is a Hermitian element of A and $\sigma(x) \geqslant 0$, then there exists a Hermitian element $y \in A$ such that $x = y^*y$.*

PROOF. Consider the subalgebra $A_1 = [x]$; it is a commutative *-subalgebra of A and, by Theorem 24.2, is isomorphic to $C(\mathfrak{M})$, $\mathfrak{M} = \mathfrak{M}(A_1)$. Hence it follows that the function $y\hat{}(M) = \sqrt{x\hat{}(M)}$ continuous on $\mathfrak{M}$ is the Gelfand transform of a Hermitian element $y \in A_1$. Thus $x = y^2 = y^*y$.

26.9. COROLLARY. *Let H_A denote the set of all Hermitian elements of A; H_A is a real subspace of the space A. Then the cone $K \subset H_A$ spanned by the elements x^*x (i.e. consisting of finite sums of the form $\sum \alpha_i x_i^* x_i$ with $\alpha_i \geqslant 0$) coincides with the set $\{x \in H_A : \sigma(x) \geqslant 0\}$. Hence it follows, by Theorem 24.6, that if A_1 is any *-subalgebra of A, then the cone K_1 spanned in H_{A_1} by the elements x^*x is precisely the set*

(26.9.1) $$K_1 = K \cap H_{A_1}.$$

We now can formulate the announced theorem on the extension of positive functionals.

26.10. THEOREM. *Let A be a* **B***-algebra with unit and let A_1 be a *-subalgebra of A. Then for every positive functional f defined in A_1 there exists a positive functional F defined in A such that $F(x) = f(x)$ for all $x \in A_1$.*

PROOF. The assertion follows easily from Lemma 26.3 with $X = H_A$, $Y = H_{A_1}$, the functional f restricted to H_{A_1} and with K as in Corollary 26.9. Since for every $z \in H_A$ the spectrum $\sigma(z)$ is real, then for a sufficiently large integer N we have $Ne - z \in K$, whence $z + Ne - z \in (z+K) \cap Y \neq \emptyset$ and Lemma 26.3 applies in view of (26.9.1).

26.11. THEOREM. *Let A be a* **B***-algebra with unit. Let P_0 denote the set of all positive funcionals in A which fulfil the condition $f(e) = 1$. Then the submultiplicative norm in A satisfying condition* (24.1.1) *is uniquely determined and is given by the formula*

(26.11.1) $$\|x\| = \sup_{f \in P_0} [f(x^*x)]^{1/2}.$$

PROOF. Fix an $x \in A$. We have, by (25.8.1),

(26.11.2) $$f(x^*x) \leqslant \|x^*x\| = \|x\|^2$$

for every functional $f \in P_0$, so that $\|x\| \geqslant \sup_{f \in P_0} [f(x^*x)]^{1/2}$. On the other hand, x^*x is a Hermitian element and thus, in view of Corollary 24.5, $\|x^*x\| = \|x^*x\|_s$. Consequently there exists a multiplicative linear functional f_M in the algebra $A_1 = [x^*x]$ such that $f_M(x^*x) = \|x^*x\|_s$. We have $f_M(e) = 1$ and clearly $f_M(z) \geqslant 0$ if $\sigma_{A_1}(z) \geqslant 0$, $z \in A_1$. Thus f_M is a positive functional in A_1 and it can be extended to a positive functional F in A, by Theorem 26.10. Since $f_M(e) = 1$, so $F \in P_0$ and $F(x^*x) = \|x^*x\| = \|x\|^2$ which, together with (26.11.2), yields the asserted equality.

Now we may pass to the proof of the basic representation theorem for **B***-algebras.

26.12. THE GELFAND–NAIMARK REPRESENTATION THEOREM. *Every* **B***-algebra* *A is isometrically *-isomorphic to some *-subalgebra of an algebra $B(H)$.*

PROOF. According to Theorem 26.1 we may assume that the algebra A has a unit e. Let P_0 denote the set of all positive functionals in A such that $f(e) = 1$. To each such functional f there corresponds a cyclic representation $x \to T_x^f$ in a Hilbert space H^f, in virtue of Theorem 25.11. Let $x \to T_x$ denote the direct sum of the family of representations $x \to T_x^f$, $f \in P_0$ and let H denote the underlying Hilbert space (see Definition 25.12). We shall show that the mapping $x \to T_x$ is an isometric *-isomorphism of the algebra A into the algebra $B(H)$. We already know that this mapping is a *-homomorphism of A into $B(H)$, so it suffices to show that it is an isometry. By formula (25.11.1) we may write $f(x) = \langle T_x^f \xi_0^f, \xi_0^f \rangle_f$, where $\langle , \rangle_f$ is the scalar product in H^f and ξ_0^f is the corresponding cyclic vector. Thus, since $f(e) = 1$, $\|\xi_0^f\|_f = 1$. On the other hand (cf. Definition 25.12), we have

$$f(x^*x) = \langle T_x^f \xi_0^f, T_x^f \xi_0^f \rangle_f = \langle T_x S_f \xi_0^f, T_x S_f \xi_0^f \rangle \leqslant \|T_x\|^2 \|S_f \xi_0^f\|^2$$
$$= \|T_x\|^2 \|\xi_0^f\|_f^2 = \|T_x\|^2 \leqslant \|x\|^2;$$

hence, by (26.11.1)

$$\|x\|^2 = \sup_{f \in P_0} f(x^*x) \leqslant \|T_x\|^2 \leqslant \|x\|^2,$$

so $\|T_x\| = \|x\|$.

26.13. REMARK. For a long time it was not known whether every *-algebra with a norm satisfying the condition

$$\|x^*x\| \geqslant C\|x\|^2$$

is a **B***-algebra (cf. Theorem 24.7). Such algebras are called *Arens *-algebras*. Yood [1] has proved that an Arens *-algebra is a **B***-algebra if $C > t_0$, where $t_0 = 0.676 \ldots$ is the real root of the equation

$$4t^3 + 2t^2 - t - 1 = 0.$$

Recently Yood proved that every Arens *-algebra is **B***-algebra (or a more general theorem from which this result follows immediately; cf. Yood [2]).

§ 27. SOME CONSEQUENCES OF THE REPRESENTATION THEOREM

The isomorphism asserted by Theorem 26.12 has yet another useful property, namely it preserves positivity. To be precise, we introduce the following concepts.

27.1. Definition. An operator $T \in B(H)$ is called a *positive operator* if

$$\langle T\xi, \xi \rangle \geqslant 0$$

for all $\xi \in H$. It is not difficult to verify that every positive operator is Hermitian (self-adjoint).

27.2. Definition. For any **B***-algebra A let us write

$$P_A = \{x \in H_A : \sigma(x) \geqslant 0\},$$

where, as previously, H_A is the set of all Hermitian elements of A and $\sigma(x) \geqslant 0$ means that the spectrum consists of real non-negative numbers. The elements of the set P_A will be called *positive elements of A*.

Now we shall show that positive elements of a **B***-algebra correspond via the Gelfand–Naimark isomorphism (Theorem 26.12) to positive operators in $B(H)$.

27.3. Theorem. *Let A be a* **B***-algebra and let Φ be an *-isomorphism of A into $B(H)$* (Theorem 26.12). *Then the set ΦP_A coincides with the set of positive operators belonging to the subalgebra $\Phi A \subset B(H)$.*

Proof. Write $\Phi a = T_a$ and let $a \in P_A$. We have, by Lemma 26.8, $a = b^* b$, whence

$$\langle T_a \xi, \xi \rangle = \langle T_{b^*b} \xi, \xi \rangle = \langle T_b \xi, T_b \xi \rangle \geqslant 0,$$

i.e. T_a is a positive operator. On the other hand, if

(27.3.1) $$\langle T_a \xi, \xi \rangle \geqslant 0$$

for all $\xi \in H$, then we have

$$\langle T_a \xi, \xi \rangle = \langle \xi, T_{a^*} \xi \rangle = \overline{\langle T_{a^*} \xi, \xi \rangle}$$

whence it follows that $\langle T_{a^*} \xi, \xi \rangle$ is a real number and

$$\langle T_{a-a^*} \xi, \xi \rangle \equiv 0,$$

$\xi \in H$. Write $b = i(a-a^*)$. Clearly $b^* = b$ and $\langle T_b \xi, \xi \rangle \geqslant 0$ for all $\xi \in H$ (the last inequality is in fact an equality). Thus the bilinear form $[\xi, \eta] = \langle T_b \xi, \eta \rangle$ has all those properties which yield the Schwarz inequality, so we have

$$|\langle T_b \xi, \eta \rangle|^2 \leqslant \langle T_b \xi, \xi \rangle \langle T_b \eta, \eta \rangle \equiv 0,$$

so that $T_b = 0$ and $a = a^*$. To show that $a \in P_A$, let us write $a = h-k$ with $h, k \in P_A$, $hk = kh = 0$, as we did in the proof of Lemma 26.6. We have

$$\langle T_{h-k} \xi, \xi \rangle \geqslant 0.$$

Suppose that $k \neq 0$. Then also $k^3 \neq 0$ and, repeating the argument used above to show that $a-a^* = 0$, we conclude that $\langle T_{k^3} \xi, \xi \rangle$ cannot be identically zero. Thus there exists a vector $\xi_0 \in H$ such that

$$\langle T_{k^3} \xi_0, \xi_0 \rangle > 0.$$

Writing $\xi_1 = T_k \xi_0$ we get

$$(27.3.1) \qquad \langle T_k \xi_1, \xi_1 \rangle = \langle T_{k^3} \xi_0, \xi_0 \rangle > 0.$$

Since $T_h \xi_1 = T_{hk} \xi_0 = 0$, then

$$-\langle T_k \xi_1, \xi_1 \rangle = \langle T_{h-k} \xi_1, \xi_1 \rangle = \langle T_a \xi_1, \xi_1 \rangle < 0$$

a contradiction to (27.3.1). So we have $k = 0$ and $a = h-k \in P_A$.

We conclude this section with the following theorem which finds applications in the theory of von Neumann algebras.

27.4. THEOREM. *A* **B****-algebra A has a unit if and only if its unit ball* $K = \{x \in A: ||x|| \leqslant 1\}$ *has extremal points.*

PROOF. We first show that if A has a unit e, then e is an extremal point of K. Suppose that $e = \frac{1}{2}(x+y)$, $x, y \in K$. We are to show that $x = y = e$. Write $u = \frac{1}{2}(x^*+x)$, $v = \frac{1}{2}(y^*+y)$. We have $u = u^*$, $v = v^*$ and $e = \frac{1}{2}(u+v)$. Since $u = 2e-v$, then $uv = vu$ and the subalgebra $A_0 \subset A$ generated by u and v (including also e by definition) is a commutative **B***-algebra. In virtue of the Gelfand–Naimark Theorem 24.2 A_0 is some $C(\Omega)$. We shall show that $u\hat{}(M) \equiv v\hat{}(M) \equiv 1$ so that $u = v = e$. Indeed, we have $||u|| = ||u||_s \leqslant 1$ and $||v|| = ||v||_s \leqslant 1$. If we had $u\hat{}(M_0) < 1$, $M_0 \in \Omega$, then $e\hat{}(M_0) = \frac{1}{2}(u\hat{}(M_0) + v\hat{}(M_0)) < 1$ would hold, which is impossible. Thus $x^* = 2u-x = 2e-x$ and $x^*x = xx^*$. Now, as $e = \frac{1}{2}(x+x^*)$, we may repeat the argument with x and x^* in place of u and v. Thus we obtain $x^* = x = e$ and, by symmetry, $y^* = y = e$. So the unit e in an extremal point of K.

The proof of the converse, i.e. of the fact that if K has extremal points, then A has a unit, will be based on several lemmas.

27.5. LEMMA. *If x is an extremal point of K, then x^*x and xx^* are projections* (Definition 23.11).

PROOF. Let A_0 be the subalgebra of A generated by the Hermitian element x^*x. It is a commutative **B***-algebra, so, by the Gelfand–Naimark Theorem, it is isometrically *-isomorphic either to a $C(\Omega)$ algebra or to a $C_0(\Omega)$ algebra with Ω compact resp. locally compact Hausdorff, depending on whether A_0 has a unit or not (Theorem 24.2 and Corollary 24.3). In either of the two cases it is easy to construct a sequence of Hermitian elements (i.e. real-valued functions on Ω) $y_n \in A_0$ such that $y_n x^*x \to x^*x$ and $y_n^2 x^*x \to x^*x$. The element x^*x will be shown to be a projection if we prove that the spectrum $\sigma(x^*x)$ consists only of the numbers 0 and 1. Since $\sigma(x^*x) \geqslant 0$ (Lemma 26.6), it suffices to show that if $t > 0$, $t \in \sigma_{A_0}(x^*x)$, then $t = 1$. We always have $t \leqslant 1$, since $t \leqslant ||x^*x||_s = ||x^*x|| = ||x||^2 = 1$. Suppose that $0 < t < 1$ and let p denote a point of Ω such that $(x^*x)\hat{}(p) = t$. Then there exists a non-negative function $a\hat{} \in A_0\hat{}$ such that $a\hat{}(p) > 0$ and $||(y_n+a)^2(x^*x)|| \leqslant 1$ and $||(y_n-a)^2(x^*x)|| \leqslant 1$. (I suffices to take a function $a^\wedge \in A_0^\wedge$ whose support is contained in the set $\{p \in \Omega: (x^*x)^\wedge(p) < \frac{1}{2}(1+t)\}$ and whose norm is lesser than $(1-\frac{1}{2}(1+t))$. Thus

$$1 \geqslant \|(y_n+a)^2x^*x\| = \|(y_n+a)x^*x(y_n+a)\|$$
$$= \|[x(y_n+a)]^*[x(y_n+a)]\| = \|x(y_n+a)\|^2,$$

so $x(y_n+a) \in K$ and similarly $x(y_n-a) \in K$. On the other hand, the sequence (y_n) has been choosen in such a way that

$$\|xy_n-x\|^2 = \|(xy_n-x)^*(xy_n-x)\| = \|x^*xy_n^2-2x^*xy_n+x^*x\| \to 0,$$

whence $x(y_n+a) \to x+xa$ and, similarly, $x(y_n-a) \to x-xa$. Therefore $x+xa \in K$ and $x-xa \in K$, whence, in view of the equality $x = \frac{1}{2}[(x+xa)+(x-xa)]$ and of the extremality of x, we obtain $x = x+xa = x-xa$. This means that $xa = 0$ and $x^*xa = 0$, which is impossible, since $(x^*x)^\wedge(p) = t > 0$ and $a^\wedge(p) > 0$. Thus x^*x is a projection. To see that xx^* is also, observe that an element $x \in K$ is extremal if and only if x^* is and repeat the argument with respect to x^*.

27.6. Lemma. *If x is an extremal point of K, then, writing*

$$B = (1-xx^*)A(1-x^*x), \tag{27.6.1}$$

(where the symbols $(1-a)b$, $b(1-a)$ stand for $b-ab$, $b-ba$), *we have* $B = \{0\}$.

Proof. Let $a \in B$. We have to show that $a = 0$. We may assume, without a loss of generality, that $\|a\| < 1$. We have

$$\|x\pm a\|^2 = \|(x\pm a)^*(x\pm a)\| = \|x^*x+a^*a\pm(x^*a+a^*x)\|. \tag{27.6.2}$$

Since $a = (1-xx^*)y(1-x^*x)$ for some $y \in A$, so, by the preceding lemma,

$$xx^*a = (xx^*-xx^*)y(1-x^*x) = 0.$$

It hence follows that $a^*xx^*a = 0$ and

$$\|x^*a\|^2 = \|(x^*a)^*(x^*a)\| = \|a^*xx^*a\| = 0,$$

whence $x^*a = 0$ and also $a^*x = (x^*a)^* = 0$. Thus (27.6.2) may be rewritten in the form

$$\|x\pm a\|^2 = \|x^*x+a^*a\|. \tag{27.6.3}$$

According to the Gelfand–Naimark Theorem 26.12 the elements x^*x and a^*a may be regarded as operators P and T in some Hilbert space H, whereby P is a projection (Lemma 27.5). Moreover, we have $PT = TP = 0$, as $a^*ax^*x = a^*(ax^*x) = a^*(1-xx^*)y(x^*x-x^*x) = 0$ and $x^*xa^*a = (ax^*x)^*a = 0$. Further, $\|P\| = 1$ and $\|T\| = 1$, as is easy to verify. Write $H_1 = PH$, $H_2 = H_1^\perp$, then every element $\xi \in H$ can be written uniquely as a sum $\xi = \xi_1+\xi_2$ with $\xi_i \in H_i$ and $P\xi_2 = 0$. Thus $T\xi_2 = (I-P)T\xi_2 \in H_2$ and $T\xi_1 = TP\xi_1 = 0$ so that for any $\xi \in H$ we have

$$\|(P+T)\xi\|^2 = \|P\xi_1+P\xi_2\|^2 = \|P\xi_1\|^2+\|T\xi_2\|^2$$
$$\leqslant \|P\|^2\|\xi_1\|^2+\|T\|^2\|\xi_2\|^2 \leqslant \|\xi_1\|^2+\|\xi_2\|^2 = \|\xi\|^2,$$

whence $\|P+T\| \leqslant 1$ and $\|x^*x+a^*a\| \leqslant 1$. From equality (27.6.3) it thus follows that $x+a \in K$ and $x-a \in K$ and since $x = \frac{1}{2}[(x+a)+(x-a)]$ is an extremal element of K, then $a = 0$.

27.7. LEMMA. *If A is any* **B***-*algebra and x, y are positive elements of A* (cf. Definition 27.2), *then* $\|x+y\| \geqslant \|x\|$.

PROOF. In virtue of Theorems 26.12 and 27.3 the elements x and y may be regarded as positive operators in some Hilbert space H: $\langle x\xi, \xi\rangle \geqslant 0$ and $\langle y\xi, \xi\rangle \geqslant 0$ for all $\xi \in H$. Thus we have for $\|\xi\| \leqslant 1$

$$\|x+y\| \geqslant \langle (x+y)\xi, \xi\rangle \geqslant \langle x\xi, \xi\rangle.$$

According to Lemma 26.8, we have $x = u^*u$ for some Hermitian u. Hence

$$\|x+y\| \geqslant \sup_{\|\xi\|\leqslant 1} \langle x\xi, \xi\rangle = \sup_{\|\xi\|\leqslant 1} \langle u\xi, u\xi\rangle = \|u\|^2 = \|u^*u\| = \|x\|.$$

27.8. LEMMA. *Let A be a* **B***-*algebra and let $x_1, \ldots, x_m \in A$. Then there exists a sequence of Hermitian elements $(d_n) \subset A$ such that $\sigma(d_n) \geqslant 0$ for $n = 1, 2, \ldots$, and the limits*

$$\lim x_k d_n = \lim_n d_n x_k = x_k$$

exist for every x_k, $k = 1, \ldots, m$.

PROOF. Let $h = x_1^*x_1 + \ldots + x_m^*x_m + x_{m+1}^*x_{m+1} + \ldots + x_{2m}^*x_{2m}$, where $x_{m+i} = x_i^*$, $i = 1, \ldots, m$. The element h is evidently Hermitian and $\sigma(h) \geqslant 0$, by Lemmas 26.4 and 26.6. Consider the subalgebra $A_0 \subset A$ generated by h. According to Theorem 24.2 and Corollary 24.3 A_0 is a $C(\Omega)$ or a $C_0(\Omega)$ algebra, depending on whether A has a unit or not. In both cases the function $d_n^\wedge = \dfrac{nh^\wedge}{1+nh^\wedge}$ is the Gelfand transform of some uniquely determined element $d_n \in A_0$. We shall show that the sequence (d_n) has the asserted property. According to the convention made in Lemma 27.6 we may write $(1-d_n)h(1-d_n) \in A_0$. We have also

$$\|(1-d_n)h(1-d_n)\| = \sup_{p\in\Omega} \frac{1}{n}\,\frac{nh^\wedge(p)}{[1+nh^\wedge(p)]^2} \leqslant \frac{1}{4n},$$

since the function $t/(1+t)^2$ does not exceed $\frac{1}{4}$ for $t \geqslant 0$. Further,

$$\begin{aligned}(1-d_n)h(1-d_n) &= (1-d_n)\sum_{i=1}^{2n} x_i^*x_i(1-d_n)\\ &= \sum_{i=1}^{2n} (1-d_n)x_i^*x_i(1-d_n) = u_k+v_k,\end{aligned}$$

where

$$u_k = \sum_{i\neq k} (1-d_n)x_i^*x_i(1-d_n) = \sum_{i\neq k} [x_i(1-d_n)]^*[x_i(1-d_n)],$$

$$v_k = (1-d_n)x_k^*x_k(1-d_n) = [x_k(1-d_n)]^*[x_k(1-d_n)].$$

The elements u_k and v_k need not belong to A_0. They are Hermitian elements and, in view of Lemmas 26.4 and 26.6, $\sigma(u_k) \geqslant 0$ and $\sigma(v_k) \geqslant 0$, $k = 1, \ldots, 2m$. Applying the preceding lemma we get

$$\|x_k(1-d_n)\|^2 = \|[x_k(1-d_n)]^*[x_k(1-d_n)]\| = \|v_k\| \leqslant \|u_k+v_k\|$$
$$= \|(1-d_n)h(1-d_n)\| \leqslant 1/4n$$

for $k = 1, \ldots, 2m$. It hence follows that $\lim_n x_k d_n = x_k$ and $\lim_n x_k^* d_n = x_k^*$ for $k = 1, \ldots, m$. The last equality yields $\lim_n (x_k^* d_n)^* = \lim_n d_n x_k = x_k$.

Now we can return to the proof of Theorem 27.4. We have to prove that the algebra A has a unit if the ball K has extremal points. Let x_0 be an extremal point of K. We shall show that for any system of elements $x_1, \ldots, x_m \in A$ the sequence (d_n) constructed in Lemma 27.8 (applied to the points $x_0, x_1, \ldots, x_m$) converges in A. In fact, suppose that this is not the case. Then there exists an increasing sequence of integers (k_n) and a $\delta > 0$ such that $\|d_n - d_{k_n}\| \geqslant \delta > 0$. Writing $z_n = (d_n - d_{k_n})/\|d_n - d_{k_n}\|$ we have $\|z_n\| = 1$ and $\lim z_n x_0 = \lim x_0 z_n = \lim x_0^* z_n = \lim z_n x_0^* = 0$, whence $\lim \|y_n\| = 1$, where

$$y_n = (1-x_0 x_0^*) z_n (1-x_0^* x_0) = z_n - x_0 x_0^* z_n - z_n x_0^* x_0 + x_0 x_0^* z_n x_0^* x_0 .$$

However, this is impossible, since the elements y_n belong to the set B defined by (27.6.1) and so, by Lemma 27.6, are all equal zero.

Finally we show that the limit of the sequence (d_n) does not depend on the choice of the points $x_1, \ldots, x_m$. Indeed, if we had $\lim d_n = e_1$ and $\lim d_n' = e_2 \neq e_1$ for the sequence (d_n') corresponding to some other couple of points $x_0, y_1, \ldots, y_k$, then putting $z = e_1 - e_2$ we would obtain $z x_0 = x_0 z = z x_0^* = x_0^* z = 0$ and so again

$$z = z - x_0 x_0^* z - z x_0^* x_0 + x_0 x_0^* z x_0^* x_0 \in B,$$

the set B being defined by (27.6.1); hence, by Lemma 27.6, $z = 0$.

Putting now $e = \lim d_n$ we obtain the unit of the algebra A.

§ 28. IRREDUCIBLE REPRESENTATIONS AND NON-DECOMPOSABLE POSITIVE FUNCTIONALS

In this section we point out some further properties of algebras with an involution. We shall prove that if an *-algebra has any representations at all, then it has also irreducible ones. More precisely, we shall show that irreducible representations correspond to the so-called non-decomposable positive functionals and we shall prove the existence of such functionals under the assumption that the algebra in question has at least one non-zero positive functional. Further we shall discuss yet some other properties of positive functionals in commutative *-algebras and we shall give an example of a commutative *-algebra which has no non-zero representation, or, equivalently, has no non-zero positive functionals.

28.1. Definition. Let $P(A)$ denote the set of all positive functionals of a given *-algebra A. We define an *ordering in A*. We shall write $f_1 \prec f_2, f_1, f_2 \in P(A)$ if there exists a positive number λ such that

$$\lambda f_2 - f_1 \in P(A).$$

28.1.a. Exercise. Show that the relation $\prec$ is indeed an ordering in $P(A)$, i.e. prove its transitivity.

28.2. Theorem. *Let $f \in P(A)$ and let $x \to T_x$ be the cyclic representation of the *-algebra A in a space H corresponding, by Theorem* 25.11, *to the functional f. Let ξ_0 denote the cyclic vector of this representation. Then every positive operator $T \in B(H)$* (see Definition 27.1) *which commutes with all operators T_x defines a positive functional f_1 given by*

(28.2.1) $$f_1(x) = \langle T_x T\xi_0, \xi_0\rangle,$$

such that $f_1 \prec f$.

Conversely, *if $f_1 \in P(A)$ is a functional fulfilling the condition $f_1 \prec f$, then there exists a positive operator $T \in B(H)$ which commutes with all T_x's and such that the functional f_1 is then given by formula* (28.2.1).

Proof. Let f_1 be the functional given by (28.2.1). We have

$$f_1(x^*x) = \langle T_{x^*x} T\xi_0, \xi_0\rangle = \langle TT_x\xi_0, T_x\xi_0\rangle \geqslant 0,$$

so $f_1 \in P(A)$. Further,

$$\langle T\xi, \xi\rangle \leqslant ||T|| \, ||\xi||^2 = ||T|| \langle \xi, \xi\rangle$$

for any $\xi \in H$, thus putting $\xi = T_x\xi_0$, we get

$$||T|| \langle T_{x^*x}\xi_0, \xi_0\rangle - \langle T_{x^*x} T\xi_0, \xi_0\rangle \geqslant 0,$$

since T commutes with all T_x's. In other words, we have

$$||T|| f(x^*x) - f_1(x^*x) \geqslant 0,$$

or

$$||T|| f - f_1 \in P(A),$$

which means that $f_1 \prec f$.

Now suppose that $f_1 \prec f$ and $f(x) = \langle T_x\xi_0, \xi_0\rangle$, $T_x \in B(H)$, $\xi_0 \in H$. In the proof of Theorem 25.11 the space H has been constructed as the completion of the quotient space resulting under an identification of those elements which are not separated by the pseudonorm $||x|| = [f(x^*x)]^{1/2}$. Since $f_1(x^*x) \leqslant \lambda f(x^*x)$ the positive functional f_1 defines a Hermitian form $\langle \xi, \eta\rangle_1$ in H whose value at $\xi = T_x\xi_0$, $\eta = T_y\xi_0$ is $f_1(y^*x)$ and which fulfils the condition $\langle \xi, \xi\rangle_1 \leqslant \lambda\langle \xi, \xi\rangle$ for $\xi \in H$. The form $\langle \xi, \eta\rangle_1$ is thus a continuous form in H and, for a fixed $\eta_1 \in H$, $\langle \xi, \eta_1\rangle_1$ is a continuous linear functional in H. Thus there exists an element $\eta_2 \in H$ such that $\langle \xi, \eta_1 \rangle_1 = \langle \xi, \eta_2 \rangle$ for all $\xi \in H$. We now define $T\eta_1$ as η_2; T is an additive homogeneous operator and $\langle \xi, T\eta \rangle = \langle \xi, \eta \rangle_1$ holds for all $\xi, \eta \in H$. Interchanging the variables ξ and η, we get also $\langle T\xi, \eta \rangle = \langle \xi, \eta \rangle_1$. So we have

$$||T||^2 = \sup_{||\xi||, ||\eta|| \leqslant 1} |\langle T\xi, \eta \rangle|^2 = \sup_{||\xi||, ||\eta|| \leqslant 1} |\langle \xi, \eta \rangle_1|^2$$
$$\leqslant \sup_{||\xi||, ||\eta|| \leqslant 1} \langle \xi, \xi \rangle_1 \langle \eta, \eta \rangle_1 \leqslant \lambda^2.$$

T is thus a bounded operator. Clearly, it is a positive operator, thus also Hermitian. It remains to show that the equality $TT_x = T_x T$ holds for all $x \in A$. It follows from the construction of H and from the definition of the form $\langle , \rangle_1$ that

$$f_1(x) = \langle TT_x \xi_0, \xi_0 \rangle,$$

whereby we are to show that

$$\langle TT_a \xi, \eta \rangle = \langle T_a T\xi, \eta \rangle = \langle T\xi, T_a \eta \rangle \tag{28.2.2}$$

for any $a \in A$ and $\xi, \eta \in H$. The operator T being continuous, it suffices to prove (28.2.2) for all ξ and η belonging to the set $\{T_x \xi_0 : x \in A\}$, dense in H. The equality of the outer terms in (28.2.2) can be now rewritten as

$$f_1(y^* a x) = f_1[(a^* y)^* x],$$

which evidently holds for all $x, y, a \in A$.

28.3. DEFINITION. A functional $f \in P(A)$ is called a *non-decomposable functional* if the conditions $f_1 \prec f, f_1 \in P(A)$ imply $f_1 = \lambda f, \lambda \geqslant 0$.

28.4. THEOREM. *A cyclic representation $x \to T_x$ corresponding to a functional $f \in P(A)$ is irreducible if and only if f is non-decomposable.*

PROOF. In virtue of Theorem 25.5, a representation $x \to T_x$ is irreducible if and only if the only operators commuting with all T_x's are the operators λI. We may here restrict ourselves to positive operators only since in the case when the representation is not irreducible there exists a projection operator, thus a positive one, commuting with all T_x's. Now the assertion follows easily from the preceding theorem.

Theorem 28.4 reduces the proof of the existence of irreducible representations of an *-algebra to that of the existence of non-decomposable positive functionals on it.

Let us consider the set $P_0(A) = \{f \in P(A) : f(e) = 1\}$. $P_0(A)$ is a closed convex subset of the conjugate space A'. It is not difficult to verify that it is A-weak compact, thus, by the Krein–Milman theorem, $P_0(A)$ has extremal elements (i.e. such which are not interior points of any segment contained in this set), provided $P_0(A)$ is not empty. We shall show that each such extremal element is a non-decomposable functional, or, equivalently, that if f is a decomposable positive functional, then f is not an extremal point of the set $P_0(A)$. Let f be decomposable and let $f_1 \prec f$, $f_1 \neq Cf$. Then there exists a $\lambda > 0$ such that $f_2 = \lambda f - f_1 \in P(A)$. Write $\alpha = f_1(e)$, $\beta = f_2(e)$. It follows from our assumption and from inequality (25.8.1) that $\alpha, \beta > 0$. So there exist elements $\tilde{f}_1, \tilde{f}_2 \in P_0(A)$ such that $f_1 = \alpha\tilde{f}_1$, $f_2 = \beta\tilde{f}_2$. Thus we have $\lambda f = \alpha\tilde{f}_1 + \beta\tilde{f}_2$, hence f is not an extremal point of $P_0(A)$, since $\frac{\alpha}{\lambda}, \frac{\beta}{\lambda} > 0$ and

$$1 = f(e) = \frac{\alpha}{\lambda}\tilde{f}_1(e) + \frac{\beta}{\lambda}\tilde{f}_2(e) = \frac{\alpha}{\lambda} + \frac{\beta}{\lambda}.$$

We have thus proved the following

28.5. THEOREM. *If an *-algebra with unit has non-zero representations, then it has also irreducible representations.*

An example of an *-algebra which has no non-zero representations at all will be given below (Remark 28.11); now we shall be concerned with positive functionals in commutative *-algebras with unit.

So we assume that A is a commutative *-algebra with unit and that the set $P(A)$ (thus also $P_0(A)$) of non-zero positive functionals is non-void.

28.6. THEOREM. *The functional $||x||_0$ defined by the formula*

$$||x||_0 = \sup_{f \in P_0} [f(x^*x)]^{1/2} \tag{28.6.1}$$

is a continuous submultiplicative pseudonorm in A satisfying the conditions $||e||_0 = 1$ and $||x^*x||_0 = ||x||_0^2$.

PROOF. From inequality (25.8.1) we get $||x||_0^2 \leqslant ||x^*x|| \leqslant ||x^*||\,||x|| = ||x||^2$ whence $||x||_0 \leqslant ||x||$. The functional $||x||_0$ is clearly homogeneous and satisfies $||e||_0 = 1$. We have to prove that it is subadditive and submultiplicative. Subadditivity follows from (25.6.2), since we have for any $f \in P_0(A)$

$$f((x^*+y^*)(x+y)) \leqslant f(x^*x) + f(y^*y) + 2[f(x^*x)f(y^*y)]^{1/2}$$
$$= ([f(x^*x)]^{1/2} + [f(y^*y)]^{1/2})^2$$

and taking on both sides the supremum over all $f \in P_0(A)$ we obtain

$$||x+y||_0 \leqslant ||x||_0 + ||y||_0.$$

Hence and from the inequality $||x||_0 \leqslant ||x||$ it follows that $||x||_0$ is a continuous pseudonorm. Concerning the submultiplicativity, note first that for any $f \in P(A)$ we have

$$|f(x)| \leqslant f(e)\,||x||_0. \tag{28.6.2}$$

In fact, putting in (25.6.2) $y = e$ we get

$$|f(x)|^2 \leqslant f(e)^2 f(x^*x)/f(e)$$

and since $f/f(e) \in P_0(A)$, then taking (on the right-hand side) the supremum over $f \in P(A)$ and passing to the square roots we obtain (28.6.2). Now, fix $y \in A$ and

$f \in P_0(A)$ and write $f_1(x) = f(y^*xy)$. It is easy to see that the functional f_1 is an element of $P(A)$. Applying (28.6.2) to this functional we obtain

$$f(y^*x^*xy) \leqslant f(y^*y)\|x^*x\|_0$$

and taking the supremum we get

$$\|xy\|_0^2 \leqslant \|y\|_0^2\,\|x^*x\|_0 \tag{28.6.3}$$

for any $x, y \in A$. Hence, putting $y = x^*$, we have $\|xx^*\|_0^2 \leqslant \|x^*\|_0^2\,\|x^*x\|_0$, and since $\|x^*x\|_0 = \|xx^*\|_0$, then

$$\|x^*x\|_0 \leqslant \|x^*\|_0^2 = \|x\|_0^2, \tag{28.6.4}$$

since clearly $\|x^*\|_0 = \|x\|_0$. From inequalities (28.6.3) and (28.6.4) it follows that

$$\|xy\|_0^2 \leqslant \|y\|_0^2\|x\|_0^2.$$

Finally, note that inequalities (28.6.3), with $y = e$, and (28.6.4) yield the desired relation $\|x^*x\|_0 = \|x\|_0^2$.

28.7. Corollary. *The set $N = \{x \in A: \|x\|_0 = 0\}$ is an ideal in the algebra A and all functionals $f \in P(A)$ are constant on the cosets of A/N. Hence the set $P(A)$ may be identified with the set $P(\tilde{A})$, where $\tilde{A}$ is the completion in the norm $\|x\|_0$ of the quotient algebra A/N. $\tilde{A}$ is a commutative* **B***-*algebra with unit and thus, by the Gelfand–Naimark theorem, $\tilde{A} = C(\mathfrak{M}(A))$. Moreover,* $\operatorname{rad} A \subset N$.

28.8. Theorem. *The space $\mathfrak{M}(\tilde{A})$, $\tilde{A}$ defined in the above Corollary, is homeomorphic to the set of all self-adjoint maximal ideals of A, i.e. such that $M^* = M$.*

Proof. The mapping $M \to M^*$ is evidently a homeomorphism of $\mathfrak{M}(A)$ onto itself, so the set $\mathfrak{M}_0$ of its fixed points is closed in $\mathfrak{M}$. We shall prove that $\mathfrak{M}_0 = \mathfrak{M}(\tilde{A})$. Let $M \in \mathfrak{M}(\tilde{A})$. The functional f_M, defined on $\tilde{A}$, may also be regarded as a multiplicative linear functional on A (constant on cosets modulo N), thus it corresponds to a maximal ideal $M' \in \mathfrak{M}(A)$. We shall show so $M'^* = M'$. Indeed, let $x \in M' \subset A$ and let $\tilde{x}^* \in \tilde{A}$ be the image of x^* under the natural homomorphism of A into A/N. $\tilde{A}$ is a commutative **B***-algebra, thus $(\tilde{x}^*)\hat{}(M) = \overline{\tilde{x}\hat{}(M)}$, whence $(\tilde{x}^*)\hat{}(M) = 0$ so that $(x^*)\hat{}(M') = 0$ and $x^* \in M'$. Hence $M'^* = M'$.

Conversely, let $M_0' \in \mathfrak{M}(A)$ and $M_0' = M_0'^*$. As can be easily verified, the multiplicative linear functional $f_{M_0'}$ is a positive functional, so, by Corollary 28.7, it is constant on the cosets modulo N and it extends to a multiplicative linear functional on $\tilde{A}$. Thus the correspondence $M \leftrightarrow M'$ between the elements of $\mathfrak{M}(\tilde{A})$ and $\mathfrak{M}_0$ is one-to-one. From the definition of topology in $\mathfrak{M}(\tilde{A})$ it follows that open sets in $\mathfrak{M}(\tilde{A})$ correspond to open sets in $\mathfrak{M}_0$, thus this correspondence is a continuous map $\mathfrak{M}_0 \to \mathfrak{M}(\tilde{A})$. Since both spaces are compact Hausdorff, this map is a homeomorphism.

As a corollary to the above considerations we obtain the following theorem.

28.9. THEOREM. *Every positive functional in A can be uniquely expressed in the form*

$$f(x) = \int_{\mathfrak{M}_0} x\hat{\ }(M)\mu(dM), \tag{28.9.1}$$

where μ is a non-negative finite measure defined on the set $\mathfrak{M}_0$ of Hermitian maximal ideals of A. Conversely, *if μ is such a measure, then formula* (28.9.1) *defines a positive functional on A.*

PROOF. We have, by Corollary 28.7 and Theorem 28.8, $\mathfrak{M}_0 = \mathfrak{M}(\tilde{A})$ and $\tilde{A} = C(\mathfrak{M}_0)$, thus, in virtue of the Riesz representation theorem, every positive functional on $\tilde{A}$ or, which in fact is the same, on A, is given by formula (28.9.1), where the measure μ is finite and is determined uniquely by the functional f. Moreover, this measure is non-negative, as can easily be seen.

The converse statement is obvious.

28.10. COROLLARY. *If A is a commutative *-algebra with unit, then a functional $f \in P(A)$ is non-decomposable if and only if it is of the form f_M with $M \in \mathfrak{M}_0$. Thus, on account of* 28.4, *the only irreducible representations of A are one-dimensional representations $x \to x\hat{\ }(M)$, where $M \in \mathfrak{M}_0$.*

PROOF. If the functional f given by (28.9.1) is non-decomposable, then the measure μ necessarily is a point-mass, since otherwise there exists a subset $\mathfrak{N} \subset \mathfrak{M}$, with $0 < \mu(\mathfrak{N}) < \mu(\mathfrak{M})$ and the functional $f_1(x) = \int_{\mathfrak{N}} x\,d\mu$ has the properties $f_1 \prec f$, $f_1 \neq \lambda f$.

28.11. REMARK. Theorem 28.9 gives a possibility of a construction of an *-algebra which has no non-zero representations or, equivalently, has no non-zero positive functionals. Namely, it suffices to construct a commutative *-algebra with unit, for which the homeomorphism $M \to M^*$ has no fixed points. The algebra $C([-1, -\frac{1}{2}] \cup [\frac{1}{2}, 1])$ with the involution $x^*(t) = \overline{x(-t)}$ has this property.

For more advanced exposition of the theory of **B***-algebras the reader is referred to Diximier [1] and Naimark [1]. We omitted here also the theory of **W***-algebras which can be found in special courses of Diximier [2], Sakai [1], Schwartz [1], or in the last chapter of Naimark [1].

COMMENTS ON CHAPTER IV

Some information about m-convex algebras with a continuous involution can be found in Brooks [2] and Tao-Hsing [2]. We quote only a few results (concerning Theorems 24.8 and 24.9).

C.24.1. THEOREM. *If A is a commutative F-algebra with unit and with a continuous involution, then every positive functional on A is continuous* (Tao-Hsing [1], [2], Żelazko [7], and Dixon [1]).

C.24.2. THEOREM. *There exists a commutative (complete) m-convex algebra with unit and with a continuous involution which has a discontinuous cyclic representation in some $B(H)$.*

For generalizations of **B***-algebras cf. Allan [2] and Dixon [1].

CHAPTER V

Function Algebras

This chapter contains elements of the theory of function algebras. This theory seems to be the branch of the theory of Banach algebras which has undergone the most vivid development; most of its results were obtained in recent years. Various applications of function algebras will not be dealt with by us because of a lack of space here; we only mention that many of them are connected with approximation theory.

In the theory of function algebras an important rôle is played by the notion of measure, so we shall first recall concepts and theorems connected with measures which will be necessary in the sequel.

Let X be a compact Hausdorff space. Every functional $f \in [C(X)]'$ can be written uniquely in the form

$$f(x) = \int_X x\, d\mu, \qquad x \in C(X),$$

where μ is a finite complex Baire measure ([1]) (the *Riesz representation theorem*). Functionals $f \in [C_R(X)]'$ are written similarly ($C_R(X)$ is the real Banach space of all continuous real-valued functions on X), the measure μ being real. Thus the conjugate space $[C(X)]'$ may be identified with the set of measures in such a way that the identification preserves linear combinations. In the sequel we shall consider only finite Baire measures defined on subsets of X.

A measure μ is said to be *concentrated* on a set $Y \subset X$ if for every measurable set $E \subset X \setminus Y$ we have $\mu(E) = 0$.

A measure μ is *non-negative* if $\int x\, d\mu \geqslant 0$ for every non-negative continuous function defined on X (we shall write $\int d\mu$ instead of $\int_X d\mu$ whenever no confusion is likely).

The *total wariation of a measure* μ over a measurable set $E \subset X$ is the number

$$|\mu|(E) = \sup \sum_{i=1}^{n} |\mu(E_i)|,$$

the supremum being taken over all possible decompositions of the set E into a finite union of disjoint measurable sets $E = \bigcup_{i=1}^{n} E_i$. The total variation of

([1]) I.e. the measure μ is defined on the σ-algebra of subsets of X generated by compact G_δ sets.

a finite Baire measure is also a finite Baire measure. The formula $||\mu|| = |\mu|(X)$ defines a norm identical with the norm of the functional corresponding to the measure μ. In particular, if a functional f can be represented by a non-negative measure μ, then $||f|| = \mu(X)$.

Every real-valued measure μ may be uniquely written as the difference $\mu = \mu_1 - \mu_2$ of two non-negative measures μ_i concentrated on disjoint subsets (the *Jordan decomposition*). In this case $|\mu| = \mu_1 + \mu_2$. From the Jordan decomposition follows the decomposition rule for complex-valued measures:

$$\mu = \mu_1 - \mu_2 + i(\mu_3 - \mu_4),$$

where $\mu_i \geqslant 0$, $i = 1, 2, 3, 4$, the decomposition being again unique if the measures μ_1, μ_2 and μ_3, μ_4 are required to be pairwise concentrated on disjoint sets.

If two measures μ_1 and μ_2 are concentrated on disjoint sets, then $||\mu_1 + \mu_2|| = ||\mu_1|| + ||\mu_2||$.

A measure μ is called a *probability measure* if $\mu \geqslant 0$ and $||\mu|| = \mu(X) = 1$. The probability measure concentrated at a single point $t \in X$ will be called the *point mass* and will be denoted by δ_t. Thus $\int x\, d\delta_t = x(t)$ for any $x \in C(X)$. It can be proved that if a measure μ is concentrated on every neighbourhood of a point $t \in X$, then it is concentrated at the point t itself and has the form $\lambda\delta_t$, with λ a complex number. It can also be shown that if the measure $|\mu|$ is concentrated at a point t, then so is μ.

A measure $\mu \geqslant 0$ is called *regular* if for any measurable set $E \subset X$ we have

$$\mu(E) = \sup_{F \in \mathscr{F}} \mu(F) = \inf_{G \in \mathscr{O}} \mu(G),$$

where $\mathscr{F}$ is the family of all measurable closed sets contained in E and $\mathscr{O}$ is the family of all measurable open sets containing E.

A complex measure μ is called *regular* if each of the four measures occurring in the Jordan decomposition of μ is regular. It can be proved that every Baire measure is regular, thus if a Baire measure μ is concentrated on a set Y, then it is concentrated also on some F_σ-subset $S \subset Y$.

A measure ν is called *absolutely continuous* with respect to a measure μ, or *μ-absolutely continuous*, if for any measurable set E the equality $\mu(E) = 0$ implies $\nu(E) = 0$.

A measure ν is called *singular* with respect to a measure μ, or *μ-singular*, if it is concentrated on a set Y of μ-measure zero, i.e. such that $|\mu|(Y) = 0$. Every measure ν may be uniquely written in the form $\nu = \nu_a + \nu_s$, where ν_a is a measure absolutely continuous with respect to a given measure μ and ν_s is μ-singular (the measures ν_a and ν_s are called the *μ-absolutely continuous* and the *μ-singular parts of ν*). If ν is a real measure, then so are ν_a and ν_s. If $\nu \geqslant 0$, then $\nu_a \geqslant 0$ and $\nu_s \geqslant 0$. If ν is a μ-absolutely continuous measure, then $\nu = \varphi\mu$, where $\varphi \in L_1(X, \mu)$, which means $\nu(E) = \int_E \varphi\, d\mu$ for all measurable $E \subset X$, or, equivalently, $\int x\, d\mu = \int \varphi x\, d\mu$

for all $x \in C(X)$ (the *Radon–Nikodym theorem*). If $\nu = \nu_a + \nu_s$ is the decomposition of ν into the absolutely continuous and the singular parts with respect to some measure μ, then $\varphi\nu = \varphi\nu_a + \varphi\nu_s$ is such a decomposition for the measure $\varphi\nu$.

If $(x_n) \subset C(X)$, $\|x_n\| \leqslant K$, $n = 1, 2, \ldots$, and $x_n(p) \to x(p)$ for μ-almost every p (i.e. for all p off a set of μ-measure zero), then $\lim \int x_n d\mu = \int x d\mu$ (*Lebesgue's theorem on the dominated convergence*). If $\int |x_n - x|^2 d\mu \to 0$, then the sequence (x_n) contains a subsequence convergent to x μ-almost everywhere (Hewitt and Stromberg [1]).

§ 29. REALIZATIONS OF FUNCTION ALGEBRAS

29.1. Definition. A commutative Banach algebra A is called a *function algebra* if its norm is equivalent to the spectral norm.

For any function algebra A without unit the algebra $A \oplus \{\lambda e\}$ also is a function algebra, so we may restrict our attention to function algebras with units. Moreover, we shall be assuming the norm in a function algebra to be identical with its spectral norm.

In the literature a function algebra is also frequently defined as any closed subalgebra (with unit) of the algebra $C(X)$, X compact Hausdorff, which separates the points of X; thus distinct points $t \in X$ correspond to different multiplicative linear functionals $x \to x(t)$. An "abstract" function algebra may admit several realizations as a subalgebra of $C(X)$, with different X's. For instance, the algebra $\mathscr{A}$ (Example 3.9) may be regarded both as a subalgebra of $C(\mathfrak{M})$, where $\mathfrak{M} = \mathfrak{M}(\mathscr{A}) = \{z \in \mathbf{C}: |z| \leqslant 1\}$ or as a subalgebra of $C(\Gamma)$, where $\Gamma = \Gamma(\mathscr{A}) = \{z \in \mathbf{C}: |z| = 1\}$. In connection therewith we give the following definition.

29.2. Definition. Let X be a compact Hausdorff space and let A be a closed separating subalgebra of $C(X)$ with unit. The pair (A, X) will be called a *concrete function algebra*; it will be also denoted by A_X. Two concrete function algebras A_X and A_Y will be regarded as identical if there exists a homeomorphism of X onto Y inducing an isomorphism between the algebras.

It is clear that every function algebra may be represented in the form of a concrete function algebra, e.g. of the algebra $(A, \mathfrak{M}(A))$. Every such representation will be called a *realization* of the function algebra A. If (A, X) is a concrete function algebra, then the mapping $t \to f_t$, where $f_t(x) = x(t)$, $t \in X$, $x \in A$ is a one-to-one continuous mapping of X into $\mathfrak{M}(A)$, i.e. a homeomorphic injection of X into $\mathfrak{M}(A)$. So we may assume that $X \subset \mathfrak{M}(A)$. On the other hand, since $\|x\| = \sup_{t \in X} |x(t)|$ for all $x \in A$, the Shilov boundary $\Gamma(A)$ is contained in X. Thus for any realization (A, X) of a function algebra A the following relation holds:

$$\Gamma(A) \subset X \subset \mathfrak{M}(A). \tag{29.2.1}$$

So if $\Gamma(A) \neq \mathfrak{M}(A)$, then the algebra A has many different realizations. Indeed, any subspace X of $\mathfrak{M}(A)$ containing $\Gamma(A)$ defines a realization, since every element

$x \in A$ is uniquely determined by its restriction to X and so every function $x|_\Gamma$, $x \in A$ extends uniquely to a function on $\mathfrak{M}$.

29.3. REMARK. A function algebra might be defined as a (closed in the sup norm) subalgebra of the algebra $C_B(X)$ of all continuous bounded functions defined on a topological space X. An examination of such algebras leads e.g. to the construction of the Čech–Stone compactification of X defined as $\beta X = \mathfrak{M}[C_B(X)]$. This type of problems will not be dealt with in this book.

The following result shows that the assumption of commutativity in the definition of a function algebra is superfluous. Here we limit ourselves to the case when the algebra possesses a unit. The theorem is, however, true also without this assumption (cf. Hirschfeld and Żelazko [1], and Le Page [1]).

29.4. THEOREM. *Let $A \in \mathbf{B}_e$ and suppose that there exists a constant C such that*

$$\|x\| \leqslant C\varrho(x) \tag{29.4.1}$$

for all $x \in A$, where $\varrho(x)$ is the spectral radius of an element $x \in A$ (cf. Definition 12.6); *then A is a commutative algebra.*

PROOF. Fix two elements $x, y \in A$ and consider the map φ: $\lambda \to \exp(-\lambda x) \times \times \exp y \exp \lambda x$ of the complex plane into A. We have, by (29.4.1),

$$\begin{aligned} |\varphi(\lambda)| &\leqslant C\varrho(\varphi(\lambda)) \\ &= C\varrho(\exp(-\lambda x)\exp y \exp \lambda x) = C\varrho(\exp y), \end{aligned} \tag{29.4.2}$$

since, by Theorem 17.7, $\varrho(uv) = \varrho(vu)$ for any $u, v \in A$. For any linear functional $f \in A'$ the map $\lambda \to f(\varphi(\lambda))$ is an entire function (cf. Exercise 18.2.a), and so, by (29.4.2) and Liouville theorem, it is a constant function, which implies that $\varphi(\lambda)$ is a constant element of A not depending on λ. This implies $\varphi(0) = \varphi(1)$ or $\exp y = \exp(-x)\exp y \exp x$, i.e.

$$\exp x \exp y = \exp y \exp x \tag{29.4.3}$$

for any $x, y \in A$. From Exercise 7.2.b it follows that if $u, v \in A$ and $\|u\|$, $\|v\| < 1$, then $u+e = \exp x$ and $v+e = \exp y$ for some $x, y \in A$, and so, by (29.4.3), $(u+e)(v+e) = (v+e)(u+e)$ or $uv = vu$. This means that all elements in A are commuting.

29.5. DEFINITION. Let (A, X) be a concrete function algebra. A functional $f \in A'$ will be called a *positive functional* if for every function $x \in A$, $x(t) \geqslant 0$ for $t \in X$, we have $f(x) \geqslant 0$.

In the case when the algebra A is symmetric, i.e. together with any function $x \in A$ also the conjugate function $\bar{x}$ is in A, and if the involution in A is defined as the conjugation $x^* = \bar{x}$, then the above definition coincides with Definition 25.6.

It will be useful for us to distinguish a smaller class of positive functionals. This class (defined for an "abstract" function algebra) consists entirely of functionals which are positive under each realization and contains all multiplicative linear functionals.

29.6. Definition. Let A be a function algebra (or a subspace of a function algebra, containing the unit). The set

$$S(A) = \{f \in A' : ||f|| = 1, f(e) = 1\}$$

will be called the *set of states of the algebra* (*space*) A (the expression being borrowed from quantum mechanics and accepted to some extent in the literature).

It can be easily verified that $S(A)$ is a weak-compact convex subset of the space A^*, so, by the Krein–Milman theorem, it is the closed convex hull of the set of its extremal points, which will in the sequel be denoted by $\Omega(A)$. The elements of the set $\Omega(A)$ will be called the *pure states* of A.

It follows from this definition that all multiplicative linear functionals are states; the following theorem shows that all states are positive functionals.

29.7. Theorem. *Let (A, X) be any realization of a function algebra A. Then every element $f \in S(A)$ may be written in the form*

(29.7.1) $$f(x) = \int x \, d\mu$$

where μ is a probability measure on X and the integration is performed over X. In particular, every functional $f \in S(A)$ is positive in every realization (A, X).

Proof. We have $A_X \subset C(X)$ and the functional $f \in S(A)$ may be extended to a functional on $C(X)$, preserving the norm. We shall denote this extension also by f. It belongs, of course, also to $S[C(X)]$, so it suffices to prove the theorem in the case $A = C(X)$. According to the Riesz representation theorem the functional f may be written in the form (29.7.1); it remains to show that the measure μ is a probability measure. Let $x \in C(X)$, $x \geqslant 0$. Let D denote any disc in $\mathbf{C}$ containing the image $x(X)$. If α is the centre and r is the radius of this disc, then $|x(t)-\alpha| \leqslant r$ for any $t \in X$, so

$$r \geqslant ||x-\alpha e|| = ||f|| \, ||x-\alpha e|| \geqslant |f(x)-\alpha f(e)| = |f(x)-\alpha|,$$

thus $f(x) \in D$. It hence follows that $f(x)$ belongs to every disc containing the set $x(X)$ and consequently also to the convex hull of this set. Thus, $x(X)$ being contained in the positive real half-axis in $\mathbf{C}$, the point $f(x)$ lies there, too, so that f is a positive functional in $C(X)$. The measure μ is thus positive and $1 = ||f|| = \mu(X)$, so μ is a probability measure.

29.8. Theorem. *Every element $f \in \Omega(A)$ is a multiplicative linear functional of A.*

Proof. Let $f \in \Omega(A)$ and let A_X be a realization of A. Write

$$Z_f = \{F \in S[C(X)] : F|_A = f\}.$$

The set Z_f is non-void by the preceding theorem; it is also convex and weak-closed (i.e. weak-compact), as can easily be seen, so it has extremal points. Let

F_0 be an extremal point of Z_f; we shall show that it is also an extremal point of the set $S[C(X)]$. In fact, suppose that $F_0 = \frac{1}{2}(F_1+F_2)$, $F_i \in S[C(X)]$. Restricting the occurring functionals to the subalgebra A we obtain $f = \frac{1}{2}(f_1+f_2)$, $f_i \in S(A)$, $f_i = F_i|_A$. By the extremality of f, $f_1 = f_2 = f$, whence $F_i \in Z_f$, and by the extremality of F_0 in Z_f we have $F_1 = F_2 = F_0$, i.e. F_0 is extremal in $S[C(X)]$. It hence follows that F_0 is a non-decomposable positive functional in the *-algebra $C(X)$ (see the proof of Theorem 28.5, the set $P_0(A)$ occurring there is precisely the set $S(A)$ for the algebra $A = C(X)$). Thus, by Corollary 28.10, F_0 is a multiplicative linear functional in $C(X)$, and so is the functional f, as the restriction of F_0 to A.

We thus see that every extremal point of the unit ball in the conjugate space A' is a multiplicative linear functional, provided it fulfils the condition $f(e) = 1$. Now we show that every extremal element of this ball has the form λf with $f \in \mathfrak{M}(A)$ and $|\lambda| = 1$. The proof will be preceded by two lemmas.

29.9. LEMMA. *Let A be a function algebra and (A, X) a concrete realization of this algebra. Let f be an extremal point of the unit ball in the conjugate space A'. Then the functional f extends to a functional F on $C(X)$, F being an extremal point of the unit ball in the space $[C(X)]'$.*

PROOF. Write $\mathscr{F} = \{F \in [C(X)]' \colon F|_A = f,\ ||F|| \leqslant 1\}$. The set $\mathscr{F}$ is non-void by the Hahn–Banach theorem. It is convex and weak-closed, so, by the Krein–Milman theorem, it has extremal points. Let F be an extremal point of $\mathscr{F}$. We shall show that F is also an extremal point of the unit ball in the space $[C(X)]$, and thus is the desired extension. Suppose that $F = \frac{1}{2}(F_1+F_2)$, $||F_i|| \leqslant 1$. By restricting both sides to A we get $f = \frac{1}{2}(f_1+f_2)$, where $f_i = F_i|_A$. We have also $||f_i|| \leqslant 1$ whence, by the extremality of f in the unit ball of A', we get $f_1 = f_2 = f$ and $F_1, F_2 \in \mathscr{F}$. Hence $F_1 = F_2 = F$ and F is an extremal point of the unit ball in $[C(X)]'$.

29.10. LEMMA. *If f is an extremal point of the unit ball of the space $[C(X)]''$ then f is a functional of the form*

$$f(x) = \lambda x(t)$$

with $|\lambda| = 1$ and $t \in X$.

PROOF. The functional f has the form $f(x) = \int x\,d\mu$ with $||f|| = |\mu|(X) = 1$. It must be shown that the measure μ is concentrated at a single point, or, which is the same, that the measure $|\mu|$ is concentrated at a single point. Note first that there exists a point $p \in X$ whose every measurable neighbourhood is of positive $|\mu|$-measure. For otherwise, if every point $t \in X$ had a neighbourhood of $|\mu|$-measure zero, then the space X could be covered by finitely many such neighbourhoods $U_1, \ldots, U_k$, resulting in

$$|\mu|(X) \leqslant \sum_{i=1}^{k} |\mu|(U_i) = 0,$$

a contradiction. Now we shall show that for any measurable neighbourhood U of p we have $|\mu|(U) = 1$ and so the measure $|\mu|$ is the point mass δ_p. Suppose, on the contrary, that $r = |\mu|(U) < 1$ for some neighbourhood U of p. Then $r > 0$ and $|\mu|(X \setminus U) = 1 - r > 0$. Define measures μ_1 and μ_2 by

$$\mu_1(E) = \mu(E \cap U), \qquad \mu_2(E) = \mu(E \setminus U)$$

for every measurable E. Note that

$$\int x\,d\mu_1 = \int_U x\,d\mu, \quad \int x\,d\mu_2 = \int_{X \setminus U} x\,d\mu.$$

We have further $\|\mu_1\| = |\mu_1|(X) = |\mu|(U) = r$ and similarly $\|\mu_2\| = 1-r$. Now put

$$f_1(x) = \frac{1}{r}\int x\,d\mu_1, \quad f_2(x) = \frac{1}{1-r}\int x\,d\mu_2.$$

Obviously $f_1 \neq f \neq f_2$ and $\|f_1\| = \|f_2\| = 1$. Moreover,

$$rf_1(x) + (1-r)f_2(x) = \int_U x\,d\mu + \int_{X \setminus U} x\,d\mu = \int_X x\,d\mu = f(x),$$

a contradiction to the supposition that f is an extremal point of the unit ball of the space $[C(X)]'$. Thus $|\mu| = \delta_p$ and consequently $\mu = \lambda\delta_p$ with $|\lambda| = 1$, so that $f(x) = \lambda x(p)$.

We now pass to the proof of the announced theorem (in a somewhat stronger formulation).

29.11. THEOREM. *Let A be a function algebra. An element φ of the unit ball of the conjugate space A' is an extremal point of this ball if and only if it has the form*

(29.11.1) $$\varphi = \lambda f,$$

where $f \in \Omega(A)$ and $|\lambda| = 1$.

PROOF. Let φ be an extremal point of the unit ball in A'. It follows from Lemmas 29.9 and 29.10 that $\varphi = \lambda f$, where $f \in \mathfrak{M}(A) \subset S(A)$. If the functional f was not an extremal point of $S(A)$, then there would exist functionals $f_1, f_2 \in S(A)$, $f_1 \neq f \neq f_2$, such that $f = \frac{1}{2}(f_1 + f_2)$. But then $\varphi = \frac{1}{2}(\lambda f_1 + \lambda f_2)$, $\lambda f_1 \neq \varphi \neq \lambda f_2$ and $\|\lambda f_1\| = \|\lambda f_2\| = 1$, contrary to the extremality of φ. So $f \in \Omega(A)$. Conversely, suppose that $f \in \Omega(A)$. It suffices to show that f is an extremal point of the unit ball in A', for then also every element $\varphi = \lambda f$ with $|\lambda| = 1$ is such a point. Suppose that $f = \frac{1}{2}(f_1 + f_2)$, $\|f_i\| = 1$. Then $|f_i(e)| \leqslant 1$ and $\operatorname{re} f_i(e) = 1$, so $f_i(e) = 1$ and $f_i \in S(A)$, whence $f_1 = f_2 = f$. Thus f is an extremal point of the unit ball of A'.

29.12. THEOREM. *The closure $\overline{\Omega(A)}$ of $\Omega(A)$ in $\mathfrak{M}(A)$ contains the Shilov boundary $\Gamma(A)$.*

PROOF. Since $\Omega(A) \subset \mathfrak{M}(A) \subset S(A)$, then

$$\sup_{f\in\bar{\Omega}} |f(x)| \leqslant \sup_{f\in\mathfrak{M}} |f(x)| = ||x|| \leqslant \sup_{f\in S(A)} |f(x)| = \sup_{f\in\operatorname{conv}\Omega} |f(x)|$$
$$= \sup_{f\in\bar{\Omega}} |f(x)|,$$

so that $||x|| = \sup_{t\in\bar{\Omega}} |x(t)|$. Thus $\overline{\Omega(A)}$ is a maximizing set for A and consequently contains the Shilov boundary $\Gamma(A)$.

29.13. REMARK. We shall show later on (Theorem 31.5) that $\overline{\Omega(A)} = \Gamma(A)$. But, in any case, it follows from the inclusion $\Gamma(A) \subset \overline{\Omega(A)} \subset \mathfrak{M}(A)$ that the pair $(A, \overline{\Omega(A)})$ is a realization of the function algebra A. We denote this realization by A_Ω and we shall apply it in the next section to the proof of an important theorem on isometries of function algebras.

§ 30. ISOMETRIES OF FUNCTION ALGEBRAS

In this section we prove Nagasawa's theorem (Nagasawa [1]) stating that two function algebras are isomorphic if and only if they are isometric when regarded as Banach spaces.

At the outset we recall some theorems concerning adjoint operators in Banach spaces. If $T: X \to Y$ is a linear operator mapping a Banach space X into a Banach space Y, then its adjoint (dual) operator T': $Y' \to X'$ is defined by the formula $\langle Tx, f\rangle = \langle x, T'f\rangle$, $x \in X$, $f \in Y'$. If T is a linear isometry of X onto Y, then T' is a linear isometry of Y' onto X', for we have $||Tx|| = ||x||$ for all $x \in X$ and so

$$||T'f|| = \sup_{||x||\leqslant 1} |\langle x, T'f\rangle| = \sup_{||x||\leqslant 1} |\langle Tx, f\rangle| = \sup_{||y||\leqslant 1} |\langle y, f\rangle| = ||f||;$$

also every $f \in X'$ is of the form $f = T^*g$ with $g(y) = f[T^{-1}(y)]$, $y \in Y$. In particular, the isometry T' carries the unit ball of Y' onto the unit ball of X', and, consequently, it carries the set of the extremal points of the unit ball in Y' onto the analogous set in X'. Note further that $(T')^{-1} = (T^{-1})'$.

30.1. THEOREM. *Let T be a linear isometry of a function algebra A_1 onto another function algebra A_2 which maps the unit e_1 of A_1 onto the unit e_2 of A_2. Then T is an isomorphism of A_1 onto A_2.*

PROOF. T is a linear homeomorphism, so it suffices to show that $T(xy) = TxTy$ for all $x, y \in A_1$. The adjoint isometry T' maps $S(A_2)$ onto $S(A_1)$ (since it preserves the norm and the unit; cf. Definition 29.6) so it also maps $\Omega(A_2)$ onto $\Omega(A_1)$. Let $f \in \Omega(A_2)$, $x, y \in A_1$. We have

$$f(Txy) = [T'f](xy) = [T'f(x)][T'f(y)] = f(Tx)f(Ty),$$

thus the operator T maps the concrete algebra $(A_1, \overline{\Omega(A_1)})$ multiplicatively onto the concrete algebra $(A_2, \overline{\Omega(A_2)})$ (cf. Remark 29.13) and so T is a multiplicative operator.

We shall now get rid of the assumption that T maps the unit of A_1 onto that of A_2. We shall namely prove the following lemma.

30.2. LEMMA. *If there exists a linear isometry T_0 of a function algebra A_1 onto a function algebra A_2, then there exists also a linear isometry T of A_1 onto A_2 preserving the unit.*

PROOF. Consider the concrete function algebras $(A_1, \overline{\Omega(A_1)})$ and $(A_2, \overline{\Omega(A_2)})$. They will also be denoted by A_1 and A_2 in this proof. It suffices to prove the existence of T for these realizations. Since the functionals in $\Omega(A_2)$ are extremal points of the unit ball in A' (Theorem 29.11), then the adjoint isometry T' maps $\Omega(A_2)$ onto a subset of the extremal points of the unit ball in A_1' and so, by Theorem 29.11, for any $f \in \Omega(A_2)$ we have $T'f = \lambda f_1$, where $f_1 \in \Omega(A_1)$ and $|\lambda| = 1$. Thus

$$|f(T_0 e_1)| = |T_0' f(e_1)| = |f_1(e_1)| = 1$$

and the image $T_0 e_1$ is a function in A_2 whose absolute value is identically equal to one. Write $u = T_0 e_1$ and $B = \{z \in C[\overline{\Omega(A_2)}]\colon\ uz \in A_2\}$. The assertion of the lemma will follow if we show that $u^{-1} \in A_2$ for then the formula $Tx = u^{-1} T_0 x$ defines the desired isometry.

We first prove that the set of functions $B \subset C[\overline{\Omega(A_2)}]$ separates the points of $\overline{\Omega(A_2)}$. Assume that $\varphi_1, \varphi_2 \in \overline{\Omega(A_2)} \subset (C[\overline{\Omega(A_2)}])'$ and $\varphi_1(z) = \varphi_2(z)$ for all $z \in B$. Since the algebra A_2 separates the points of $\overline{\Omega(A_2)}$, then by the Stone–Weierstrass theorem, every function $x \in C[\overline{\Omega(A_2)}]$ is a limit (in the norm) of the form

$$x = \lim_k \sum_{i=1}^{n_k} \overline{x_{i,k}}\, y_{i,k} = \lim_k \sum_{i=1}^{n_k} \overline{(u^{-1} x_{i,k})}(u^{-1} y_{i,k}),$$

where $x_{i,k}, y_{i,k} \in A_2$ and thus $u^{-1}x_{i,k}, u^{-1}y_{i,k} \in B$. We have $\varphi_i(\bar{y}) = \overline{\varphi_i(y)}$. It hence follows that

$$\varphi_1(x) = \lim_k \sum_{i=1}^{n_k} \overline{\varphi_1(u^{-1}x_{i,k})}\,\varphi_1(u^{-1}y_{i,k})$$

$$= \lim_k \sum_{i=1}^{n_k} \overline{\varphi_2(u^{-1}x_{i,k})}\,\varphi_2(u^{-1}y_{i,k}) = \varphi_2(x),$$

thus $\varphi_1 = \varphi_2$ and B separates the points of $\overline{\Omega(A_2)}$.

The space B contains the unit e_2 so we may talk about the set $\Omega(B)$ of the pure states of this space (Definition 29.6). From the fact that B separates the points of $\overline{\Omega(A_2)}$ and from Lemma 29.9 it follows that every element $\varphi \in \Omega(B)$

extends uniquely to a pure state of the algebra $C[\overline{\Omega(A_2)}]$. The pure states of this algebra are precisely the points of $\overline{\Omega(A_2)}$ (since a measure representing a multiplicative linear functional in a $C(X)$ algebra is necessarily of the form δ_t, $t \in X$ and thus cannot be decomposed into a convex linear combination of other non-negative measures). After an identification of the elements of $\Omega(B)$ with their extensions we can write

$$\Omega(B) \subset \overline{\Omega(A_2)}.$$

Regarding an element $\varphi \in \Omega(A_2)$ as a multiplicative linear functional on the algebra $C[\overline{\Omega(A_2)}]$ and observing that $u^{-1} = \bar{u}$ we get

(30.2.1) $$\varphi(x) = \varphi(xu^{-1}u) = \varphi(xu^{-1})\varphi(u) = \varphi(xu^{-1})\overline{\varphi(u^{-1})}$$

for any $x \in C[\overline{\Omega(A_2)}]$, in particular, for any $x \in A_2$.

Now we shall show that the restriction $\varphi|_B$ belongs to $\Omega(B)$. For suppose that $\varphi|_B$ is not a pure state in B so that there exist functionals $\psi_1, \psi_2 \in S(B)$ different from $\varphi|_B$ and such that $2\varphi|_B = \psi_1 + \psi_2$. Let $x \in A_2$. In view of (30.2.1) we may write (note that $u^{-1}x \in B$ for $x \in A_2$):

$$\begin{aligned}4\varphi(x) &= 2\varphi(u^{-1}x)2\overline{\varphi(u^{-1})} = [\psi_1(u^{-1}x)+\psi_2(u^{-1}x)][\overline{\psi_1(u^{-1})}+\overline{\psi_2(u^{-1})}] \\ &= \psi_1(u^{-1}x)\overline{\psi_1(u^{-1})}+\psi_1(u^{-1}x)\overline{\psi_2(u^{-1})}+ \\ &\qquad +\psi_2(u^{-1}x)\overline{\psi_1(u^{-1})}+\psi_2(u^{-1}x)\overline{\psi_2(u^{-1})}.\end{aligned}$$

Now put

$$\sigma_1(x) = \psi_1(u^{-1}x)\overline{\psi_1(u^{-1})}, \quad \sigma_2(x) = \psi_1(u^{-1}x)\overline{\psi_2(u^{-1})},$$
$$\sigma_3(x) = \psi_2(u^{-1}x)\overline{\psi_1(u^{-1})}, \quad \sigma_4(x) = \psi_2(u^{-1}x)\overline{\psi_2(u^{-1})},$$

where $x \in A_2$. It is not difficult to verify that $\sigma_1, \sigma_2, \sigma_3, \sigma_4 \in S(A_2)$ and since $4\varphi = \sigma_1+\sigma_2+\sigma_3+\sigma_4$ and $\varphi \in \Omega(A_2)$, so $\sigma_1 = \sigma_2 = \sigma_3 = \sigma_4$. Hence $\psi_1(u^{-1}x) = \psi_2(u^{-1}x)$ for all $x \in A_2$, thus $\psi_1(x) = \psi_2(x)$ for all $x \in B$ and so $\varphi|_B = \psi_1 = \psi_2$, contrary to the supposition. Thus $\varphi|_B \in \Omega(B)$.

So we have $\Omega(A_2) \subset \Omega(B)$. Similarly $\Omega(B) \subset \Omega(A_2)$, since A_2 is obtained from B by an analogous operation of multiplication by u. Thus $\Omega(A_2) = \Omega(B)$. We define T as $u^{-1}T_0$. The operator T is an isometry of A_1 onto the space B and maps the unit e_1 onto $e_2 \in B$. The adjoint isometry T' thus maps $\Omega(B)$ onto $\Omega(A_1)$. Let $v, w \in B$. There exist elements $x, y \in A_1$ such that $v = Tx$ and $w = Ty$. Let $\varphi \in \Omega(B) = \Omega(A_2)$. We have $z = Txy \in B$ and

$$\varphi(z) = \varphi(Txy) = T'\varphi(xy) = T'\varphi(x)T'\varphi(y) = \varphi(Tx)\varphi(Ty) = \varphi(v)\varphi(w).$$

Thus $z(t) = v(t)w(t)$ for all $t \in \Omega(A_2)$, consequently also for all $t \in \overline{\Omega(A_2)}$, which means that the product of two functions from B is again in B and B is a subalgebra of $C[\overline{\Omega(A_2)}]$. Thus for any $x, y \in A_2$ we have $u^{-1}xu^{-1}y \in B$, so $xu^{-1}y \in A_2$. Putting $x = y = e_2$ we get $u^{-1} \in A_2$.

30.3. LEMMA. *Any two isomorphic function algebras are isometric.*

PROOF. The assertion follows immediately from the fact that an isomorphism between two function algebras preserves the spectra of elements, thus also the spectral norm, thus also the norm.

From Lemmas 30.2 and 30.3 and from Theorem 30.1 we obtain

30.4. NAGASAWA'S THEOREM [1]. *Two function algebras are isomorphic if and only if they are isometric when regarded as Banach spaces* (*with the sup norm*).

30.5. REMARK. In the above theorem the expression "an isometry of Banach spaces" stands for a linear isometry. This is an essential observation, since the Mazur–Ulam theorem (Mazur and Ulam [1]), stating that every isometry of real Banach spaces preserving the origin is linear, is not valid in complex spaces.

§ 31. THE BISHOP BOUNDARY AND THE CHOQUET BOUNDARY

In the section dealing with the Shilov boundary we proved that for any commutative Banach algebra A with unit there exists the minimal closed subset $\Gamma \subset \mathfrak{M}$ such that every function $|x^\wedge|$, $x^\wedge \in A^\wedge$ attains its maximum on Γ. A natural question arises whether an analogous minimal set exists if no closedness is required. In connection therewith we give the following definition.

31.1. DEFINITION. The *Bishop boundary* of a commutative Banach algebra A with unit is the smallest subset $\Gamma_B = \Gamma_B(A) \subset \mathfrak{M}$ on which every Gelfand transform $x^\wedge \in A^\wedge$ attains its maximal absolute value, provided such a set exists ("the smallest" means that this set should be contained in every other set with this property).

The equality $\overline{\Gamma_B(A)} = \Gamma(A)$ clearly holds for any algebra which has the Bishop boundary. If A is a function algebra and (A, X) is any of its realizations, then the inclusion

(31.1.1) $$\Gamma_B(A) \subset X$$

holds by (29.2.1), if A has the Bishop boundary.

31.1.a. EXERCISE. Let $A = \{x \in \mathscr{A}: x(0) = x(1)\}$, where $\mathscr{A}$ is the algebra of all functions holomorphic in the open disc and continuous in the closed disc $K = \{z \in \mathbf{C}: |z| \leqslant 1\}$. Show that the space $\mathfrak{M}(A)$ is obtained from the disc K by the identification of the points 0 and 1. Denote by 1 the point obtained by this identification. Prove that $\Gamma(A) = \{z: |z| = 1\}$ and $\Gamma_B(A) = \Gamma(A) \setminus \{1\}$, so that $\Gamma_B(A) \neq \Gamma(A)$.

31.1.b. EXERCISE. Let X denote a product of uncountably many copies of the segment $[0, 1]$; thus every point $t \in X$ has uncountably many coordinates t_α. Let

$$A = \{x \in C(X): x(t) \text{ depends on countably many of the coordinates of } t \text{ only}\}.$$

A is a separating symmetric closed subalgebra of the algebra $C(X)$ and so $A = C(X)$ by the Stone–Weierstrass theorem. Write

$$N_0 = \{t \in X: t_\alpha \neq 0 \text{ for at most countably many indices}\},$$

$$N_1 = \{t \in X: t_\alpha \neq 1 \text{ for at most countably many indices}\}.$$

Prove that every function in $C(X)$ attains its maximal absolute value both on N_0 and N_1. Conclude hence that the algebra $A = C(X)$ does not have the Bishop boundary.

The space X in Exercise 31.1.b is non-metrizable. This is not casual; we shall prove below that if a function algebra A has a realization (A, X) with X metrizable, then A has the Bishop boundary. We first give the following definition.

31.2. DEFINITION. Let A be a commutative Banach algebra with unit. A *peak point of the space* $\mathfrak{M}(A)$ (or *of the algebra* A) is any point $M_0 \in \mathfrak{M}$ for which there exists an element $x \in A$ such that $|x^\wedge(M_0)| > |x^\wedge(M)|$ for all $M \in \mathfrak{M}(A)$, $M \neq M_0$ (this is a particular case of a peak set, see Definition 18.9).

Of course, all peak points are elements of the Shilov boundary $\Gamma(A)$, thus if A is a function algebra and A_X is any of its realizations, then all its peak points belong to X.

We now formulate the theorem referred to a few lines above. This theorem is due to Bishop [1].

31.3. THEOREM. *If a function algebra A has a realization A_X, where X is a metrizable space, then the algebra A has the Bishop boundary Γ_B which is precisely the set of all peak points of A.*

PROOF. Let $P \subset X$ denote the set of all peak points of A and let $M(x) = \{t \in X: |x(t)| = \|x\|\}$ for $x \in A \subset C(X)$. If $N \subset X$ is any maximizing set for A (i.e. such that every function $|x|$, $x \in A$ attains its maximum on N), then $P \subset N$. So it suffices to show that the set P is itself maximizing or, which is the same, that $P \cap M(x) \neq \emptyset$ for all $x \in A$. So let us fix an element $x \in A$ and consider the class $\{F_\alpha\}$ of families of subsets of X such that

(i) Every family F_α consists of sets $M(z)$, $z \in A$ and contains the element $M(x)$;

(ii) Every finite subfamily of each F_α has non-void intersection.

The class $\{F_\alpha\}$ is non-void and is ordered by the inclusion. It is not difficult to see that this class fulfils the conditions of the Kuratowski–Zorn lemma and thus contains a maximal element. Let F_0 be such a maximal element. It has the property that every larger family of sets $M(z)$ contains finitely many elements $M(z_1), \ldots, M(z_k)$ whose intersection is void. It follows, by the compactness, of the space X and of the elements of F_0, that the intersection

$$D = \bigcap F_0 = \bigcap_{M(z) \in F_0} M(z) \subset X$$

is non-void and compact. The space X being metrizable has a countable base and since the sets $X \setminus M(u)$, $M(u) \in F_0$ cover $X \setminus D$, so there exists a sequence $(u_n) \subset A$ such that $M(u_n) \in F_0$ and the sets $X \setminus M(u_n)$ cover $X \setminus D$. Thus we have $D = \bigcap_{n=1}^{\infty} M(u_n)$. Fix a point $t_0 \in D$ and write

$$x_n = [u_n(t_0)]^{-1} u_n.$$

We have, clearly, $M(x_n) = M(u_n)$ and $||x_n|| = x_n(t_0) = 1$, so the series $y = \sum_{n=1}^{\infty} 2^{-n}x_n$ converges in A and $||y|| = y(t_0) = 1$. If $t \in X \setminus M(u_k)$, then

$$|y(t)| \leqslant \sum_{n=1}^{\infty} 2^{-n}|x_n(t)| < 1,$$

since $|x_n(t)| \leqslant 1$ for all n and $|x_k(t)| < 1$. Hence $M(y) \subset M(u_k)$ for all k, so

$$M(y) \subset \bigcap_{k=1}^{\infty} M(u_k) = D.$$

Now we shall show that the set $M(y)$ consists of a single point, thus necessarily a peak point. Suppose that $M(y)$ contains at least two points. Then there exists a function $v \in A$ which is non-constant on $M(y)$. We may assume that $\max_{M(y)} |v(t)| = 1$ and that v has the value 1 at some point of $M(y)$. Consider the function $\tilde{v} = v+v^2 \in A$. Its maximal absolute value is 2; this value is attained precisely at those points where the function v is equal 1. Hence the function $|\tilde{v}|$ is not constant on $M(y)$ and the set

$$E = \{t \in M(y) \colon |\tilde{v}(t)| = 2\}$$

is a closed proper subset of $M(y)$. Choose a point $t_1 \in E$ and write

$$y_0 = [y(t_1)]^{-1}y, \qquad v_0 = [\tilde{v}(t_1)]^{-1}\tilde{v}.$$

We have $||y_0|| = y_0(t_1) = 1$, $M(y_0) = M(y)$, and $v_0(t_1) = 1$, $\sup_{t \in M(y)} |v_0(t)| \leqslant 1$, $|v_0(t)| < 1$ for $t \in M(y) \setminus E$. Write further

$$K = ||v_0||, \quad V_n = \{t \in X \colon 1+2^{-n}(K-1) \leqslant |v_0(t)| \leqslant 1+2^{-n+1}(K-1)\};$$

we then have

$$\bigcup_{n=1}^{\infty} V_n = \{t \in X \colon |v_0(t)| > 1\}, \qquad V_n \cap M(y_0) = \emptyset,$$

so $|y_0(t)| < 1$ for $t \in V_n$, $n = 1, 2, \ldots$ It follows, by the compactness of V_n's, that there exist integers k_n such that $|y_0(t)|^{k_n} \leqslant \frac{1}{2}$ for $t \in V_n$; and since $||y_0|| = 1$, the series

$$z = v_0 + 4(K-1) \sum_{n=1}^{\infty} 2^{-n} y_0^{k_n}$$

converges in A. We have

$$z(t_1) = 1+4(K-1) \sum_{n=1}^{\infty} 2^{-n} = 1+4(K-1).$$

Let us calculate the value of $||z|| = \sup_t |z(t)|$. If $t \in M(y) \setminus E$, then $|v_0(t)| < 1$ and $|y_0(t)| = 1$, whence $|z(t)| < 1+4(K-1)$. If $t \in X \setminus \bigcup_{n=1}^{\infty} V_n$, then $|v_0(t)| \leqslant 1$

and $|y_0(t)| \leqslant 1$, so that $|z(t)| \leqslant 1+4(K-1)$. If $t \in V_j$, then $|v_0(t)|$ $\leqslant 1+2^{-j+1}(K-1)$, so $|y_0(t)|^{k_n} \leqslant 1$ for all n, and $|y_0(t)|^{k_j} \leqslant \frac{1}{2}$, whence

$$|z(t)| \leqslant 1+2^{-j+1}(K-1)+4(K-1)(1-2^{-j-1}) = 1+4(K-1).$$

Hence we have

$$\|z\| = 1+4(K-1) = z(t_1) \quad \text{and} \quad M(z) \cap [M(y) \setminus E] = \emptyset.$$

Since $t_1 \in M(y) \subset D = \bigcap F_0$, then $M(z) \in F_0$ by the maximality of F_c (for the intersection of any finite subfamily of F_0 and the set $M(z)$ contains the point t_1). Thus $M(y) \subset D = \bigcap F_0 \subset M(z)$ which, the set $M(y) \setminus E$ being non-void, leads to a contradiction with the fact that the set $M(y) \setminus E$ is disjoin with $M(z)$. The supposition that $M(y)$ consists of more than one point led u to a contradiction, so $M(y) = \{t_0\}$ and t_0 is a peak point. Since $M(y) \subset I$ $= \bigcap F_0 \subset M(x)$, then $t_0 \in M(x) \cap P$ and the intersection $M(x) \cap P$ is non void for any $x \in A$, which means that P is a maximizing set.

It also follows from the proof that every function algebra, which has a realiza tion A_X with metrizable X, has peak points.

We shall now discuss the concept of the Choquet boundary. We made us of this notion already in the foregoing sections without an explicit referenc to its name.

31.4. DEFINITION. The *Choquet boundary* of a function algebra A is the se $\Omega(A)$ of pure states of this algebra. So we have $\Omega(A) \subset \mathfrak{M}(A)$ (cf. Theorem 29.8).

We now prove the theorem announced in Remark 29.13.

31.5. THEOREM. *For any function algebra A the closure of its Choquet boundar $\overline{\Omega(A)}$ is identical with its Shilov boundary $\Gamma(A)$.*

PROOF. It follows from Remark 29.13 that $\Gamma(A) \subset \overline{\Omega(A)}$. On the other hand it follows from the proof of Theorem 29.8, that if (A, X) is any realization of A then $\Omega(A) \subset X$. In particular, putting $X = \Gamma(A)$ we get $\Omega(A) \subset \Gamma(A)$ whenc the assertion follows.

31.6. COROLLARY. *For any realization A_X of a function algebra A we hav $\Omega(A) \subset X$.*

Let us now consider some realization (A, X) of a function algebra A. A func tional $f \in \mathfrak{M}(A)$ can, in general, be written in many ways in the form $f(x) = \int x\,d$ with μ a probability measure on X. However, the elements of the Choquet bound ary can be written but uniquely in this form, as follows from the followin theorem.

31.7. THEOREM. *Let A be a function algebra and let (A, X) be one of its realiza tions. A multiplicative linear functional f on A belongs to the Choquet boundar* *if and only if the measure* $\mu = \delta_p$, $p \in \Omega(A)$ *is the only representing probability measure for* f.

PROOF. Let f_p be the functional corresponding to a point $p \in \Omega(A)$. We have

$$f_p(x) = \int_X x(t)\, d\mu(t),$$

where μ is a probability measure on X. We shall prove that $\mu = \delta_p$. It will be sufficient to show that for any compact set $K \subset X \setminus \{p\}$ we have $\mu(K) = 0$. Thus suppose that there exists such a set K with $\mu(K) > 0$. Then there exists a point $q_0 \in K$ such that $\mu(K \cap U) > 0$ holds for any neighbourhood U of q_0 in X. For if $\mu(K \cap U_q) = 0$ held for all $q \in K$, U_q a neighbourhood of q, then we would have $\mu(K) = 0$ by the compactness of K. Further, there exists a neighbourhood U_0 of q_0 such that $r = \mu(U_0 \cap K) < 1$, for otherwise the measure μ would be concentrated at the point q_0, so that $f_p = f_{q_0}$ and $p = q_0$, which is impossible. Now, write $V_0 = U_0 \cap K$ and

$$\mu_1(Z) = \frac{1}{r}\mu(V_0 \cap Z),$$

$$\mu_2(Z) = \frac{1}{1-r}\mu[(X \setminus V_0) \cap Z]$$

for any Borel subset Z of X. We thus obtain two different probability measures on X such that $\mu = r\mu_1 + (1-r)\mu_2$, a contradiction, since both μ_1 and μ_2 are representing measures for some states of A and μ is representing for a pure state.

Conversely, if $f \in \mathfrak{M}(A) \setminus \Omega(A)$, then there exist functionals $f_1, f_2 \in S(A)$, $f_1 \neq f \neq f_2$, such that $f = \frac{1}{2}(f_1 + f_2)$. The corresponding probability measures are thus different and their sum is representing for $2f$. This sum cannot be concentrated at a single point, since this never can happen to the sum of two probability measures, as is easy to verify. It hence follows that there exists a probability measure representing f, which is not a point mass δ_p.

It follows from Theorem 31.7 that if $f \in \mathfrak{M}(A) \setminus \Omega(A)$ and A_X is a realization of A, then the following situations can possibly occur:

(i) A probability measure μ on X representing f is uniquely determined but is not a point mass δ_p;

(ii) There exist two different probability measures on X representing f.

31.7.a. EXERCISE. Give an example of a function algebra A, a functional $f \in \mathfrak{M}(A) \setminus \Omega(A)$, and two concrete realizations A_{X_1}, A_{X_2} of A such that the situations (i) and (ii) hold respectively in A_{X_1} and A_{X_2}.

31.8. COROLLARY. *Every peak point of A belongs to* $\Omega(A)$.

PROOF. Let (A, X) be any realization of A. Let f be a peak point of A (i.e. the functional corresponding to a peak point). Then there exists an element $x \in A$

such that $f(x) = 1$ and $|x(t)| < 1$ for all t except for a single point $p \in X$. If μ is a measure on X representing the functional f, then

$$1 = f(x) = \int x\,d\mu \leqslant \int |x|\,d\mu \leqslant 1 .$$

The last integral can be equal 1 only in the case when the measure μ is concentrated at a single point $p \in X$ with $x(p) = 1$. Thus the representing measure μ is uniquely determined and has the form δ_p, so, by the preceding theorem, $f \in \Omega(A)$.

As is seen from Exercise 31.1.b, the Bishop boundary can fail to exist whereas the Choquet boundary always exists. In the case when $\Gamma(A)$ is a metrizable space the two boundaries coincide.

A reader interested in the above outlined topics is referred to the papers Bishop [1] and Phelps [1].

§ 32. DIRICHLET ALGEBRAS

32.1. Definition. Let Y be a subspace of a Banach space $C(X)$, X compact Hausdorff. A functional $f \in [C(X)]'$ is said to be *orthogonal to the subspace* Y, in symbols $f \perp Y$, if $f(x) = 0$ for every $x \in Y$. Similarly, a Baire measure μ is *orthogonal to* Y ($\mu \perp Y$) if the functional represented by this measure is orthogonal to Y.

In order to formulate the definition of a Dirichlet algebra we shall need the following theorem.

32.2. Theorem. *For every function algebra A the following conditions are equivalent:*

(32.2.1) *The only real measure μ on $\Gamma(A)$ orthogonal to A is the measure $\mu = 0$.*

(32.2.2) *Every continuous real-valued function on $\Gamma(A)$ can be uniformly approximated by the real parts of functions belonging to A.*

Proof. Suppose that the algebra A satisfies condition (32.2.2) and let $\mu \perp A$ be a real measure defined on $\Gamma(A)$. Then $\int x\,d\mu = 0$, hence also $\int \mathrm{re}\,x\,d\mu = 0$ for all $x \in A$. It follows that $\int \varphi\,d\mu = 0$ for every continuous real-valued function φ defined on $\Gamma(A)$, so $\mu = 0$. Thus condition (32.2.2) implies (32.2.1). Now suppose that condition (32.2.2) is not fulfilled, i.e. the closure Y of the set $\{\mathrm{re}\,x \colon x \in A\}$ is a proper subspace of the space $C_R(\Gamma)$ of continuous real-valued functions defined on Γ. So there exists a real functional $f \in [C_R(\Gamma)]'$ orthogonal to Y, i.e. there exists a non-zero real measure μ on $\Gamma(A)$ orthogonal to Y. Thus $\int \mathrm{re}\,x\,d\mu = 0$ for all $x \in A$ and so

$$\int x\,d\mu = \int \mathrm{re}\,x\,d\mu + i\int \mathrm{im}\,x\,d\mu = -\int \mathrm{re}\,ix\,d\mu = 0 .$$

Hence μ is a non-zero real measure orthogonal to A and condition (32.2.1) is not fulfilled.

32.3. DEFINITION. A function algebra A is called a *Dirichlet algebra* if it satisfies the (equivalent) conditions (32.2.1) and (32.2.2).

The expression "Dirichlet algebra" is due to Gleason [1] who also refers to real parts of elements of a function algebra as "harmonic functions". In this terminology a function algebra is Dirichlet if "harmonic functions" are dense in the space of all continuous real-valued functions on the Shilov boundary.

32.3.a. EXERCISE. Prove that the algebra $\mathscr{A}$ (Example 3.9) is a Dirichlet algebra.

32.4. THEOREM. *If A is a Dirichlet algebra, then $\Omega(A) = \Gamma(A)$.*

PROOF. If $\Omega(A) \neq \Gamma(A)$, then there exists a functional $f \in \Gamma \setminus \Omega$, so, by Theorem 31.7, there exist at least two distinct probability measures on Γ representing f (δ_f may be taken as one of them). The difference of these measures is then a non-zero real measure orthogonal to A, so that A does not satisfy condition (32.2.1), i.e. A is not a Dirichlet algebra.

32.4.a. EXERCISE. Prove that the function algebra $A = \{x \in \mathscr{A} : x'(0) = 0\}$ is not a Dirichlet algebra though the equality $\Omega(A) = \Gamma(A)$ holds for this algebra (since every point of $\Gamma(A)$ is a peak point and Theorem 31.8 applies). So the converse of Theorem 32.4 is not true.

REMARK. Some authors define a Dirichlet algebra as any concrete algebra (A, X) such that every continuous real-valued function on X can be uniformly approximated by real parts of elements of A. This definition is equivalent to 32.3 in the sense that if (A, X) is such a concrete Dirichlet algebra, then X is necessarily equal to $\Gamma(A)$. For otherwise there would exist (as in the proof of Theorem 32.4) a real measure on X orthogonal to A and so (as in the proof of Theorem 32.2) the functions $\operatorname{re} x$, $x \in A$, would not be dense in $C_R(x)$. Thus the only realizations of Dirichlet algebras in the above sense are algebras $(A, \Gamma(A))$.

32.5. REMARK. If $f \in S(A)$, A a Dirichlet algebra, then the probability measure on $\Gamma(A)$ representing the functional f is uniquely determined. We then shall write $\mu \in S(A)$, or $\mu \in \mathfrak{M}(A)$ if $f \in \mathfrak{M}(A)$.

We shall now be concerned with a more detailed description of the structure of measures belonging to $A^\perp \subset [C(\Gamma)]'$ (i.e. measures orthogonal to A), where A is some Dirichlet algebra.

32.6. LEMMA. *For any measure $\mu \in \mathfrak{M}(A)$ and any F_σ set $S \subset \Gamma$ of μ-measure zero there exists a sequence of elements $x_n \in A$ such that:*

(32.6.1) $$|x_n(p)| \leqslant 1 \quad \text{for } p \in \Gamma,$$

(32.6.2) $$x_n \to 1 \quad \mu\text{-almost everywhere on } \Gamma,$$

(32.6.3) $$x_n(p) \to 0 \quad \text{for every } p \in S.$$

PROOF. We may assume that $S = \bigcup_{i=1}^{\infty} K_i$, each K_i being a closed subset of Γ, $K_i \subset K_{i+1}$ for $i = 1, 2, \ldots$ It follows from the regularity of μ that there exist closed sets $F_n \subset \Gamma \setminus S$ such that

$$\mu(\Gamma \setminus F_n) \leqslant \frac{1}{2n^2}.$$

By the Urysohn lemma there exist real-valued functions $\varphi_n \in C(\Gamma)$ such that

$$\varphi_n(p) = \begin{cases} 0 & \text{for } p \in F_n \\ x-n & \text{for } p \in K_n \end{cases} \quad \text{and} \quad 0 \geqslant \varphi_n(p) \geqslant -n \quad \text{for } p \in \Gamma.$$

Further, there exist elements $y_n \in A$ such that

$$\left| \operatorname{re} y_n(p) - \left[\varphi_n(p) - \frac{1}{4n} \right] \right| \leqslant \frac{1}{4n}$$

holds for all $p \in \Gamma$ (hence we get, in particular, $\operatorname{re} y_n(p) \leqslant \varphi_n(p)$). Thus

$$0 \leqslant -\int \operatorname{re} y_n \, d\mu = \left| \int \operatorname{re} y_n \, d\mu \right| \leqslant \int \left| \operatorname{re} y_n - \left(\varphi_n - \frac{1}{4n} \right) \right| d\mu + \int_{\Gamma \setminus F_n} |\varphi_n| \, d\mu +$$

$$+ \int_{F_n} |\varphi_n| \, d\mu + \frac{1}{4n} \leqslant \frac{1}{4n} + \frac{1}{2n} + \frac{1}{4n} = \frac{1}{n},$$

or

$$0 \geqslant \int \operatorname{re} y_n \, d\mu \geqslant -\frac{1}{n}.$$

Replacing, if necessary, each element y_n by an appropriate translate of the form $y_n + ic_n e$, c_n real, we may assume that $\int \operatorname{im} y_n \, d\mu = 0$, $n = 1, 2, \ldots$

Now put $\tilde{y}_n = \exp y_n$. We have $\tilde{y}_n \in A$ and $\|\tilde{y}_n\| = \sup_{\Gamma} \exp(\operatorname{re} y_n) \leqslant 1$. If f is the multiplicative linear functional on A represented by the measure μ, then

$$1 \geqslant \int \tilde{y}_n \, d\mu = f(\tilde{y}_n) = \exp f(y_n) = \exp\left(\int \operatorname{re} y_n \, d\mu + i \int \operatorname{im} y_n \, d\mu \right)$$

$$= \exp\left(\int \operatorname{re} y_n \, d\mu \right) \geqslant \exp\left(-\frac{1}{n} \right) \to 1.$$

Hence

$$0 \leqslant \int |\tilde{y}_n - e|^2 \, d\mu = \int |\tilde{y}_n|^2 \, d\mu + 1 - 2\operatorname{re} \int \tilde{y}_n \, d\mu$$

$$\leqslant 2\left(1 - \operatorname{re} \int \tilde{y}_n \, d\mu\right) = 2\left(1 - \int y \, d\mu\right),$$

since $\int \tilde{y}_n \, d\mu$ is a real number. Hence the sequence $(\tilde{y}_n)$ converges to e in the space $L_2(\mu)$, so it contains a subsequence, which will be denoted by (x_n), μ-almost everywhere convergent to e. The sequence (x_n) satisfies conditions (32.6.1) and

(32.6.2). Condition (32.6.3) also is fulfilled, for if $p \in S$, then there exists an index n_0 such that $p \in K_{n_0} \subset K_n$ for $n \geqslant n_0$ and

$$|\tilde{y}_n(p)| = \exp(\operatorname{re} y_n(p)) \leqslant \exp(\varphi_n(p)) = \exp(-n)$$

holds for $n \geqslant n_0$.

32.7. THEOREM. *Let $\mu \in \mathfrak{M}(A)$, A a Dirichlet algebra, and let $\nu \in A^{\perp}$. Let ν_a and ν_s denote the μ-absolutely continuous and μ-singular parts of ν, respectively. Then both ν_a and ν_s belong to $A^{\perp}$.*

PROOF. There exists a set $S \subset \Gamma$ with $\mu(S) = 0$ and such that the measure ν_s is concentrated on S. We may assume that S is an F_σ set, so there exists a sequence $(x_n) \subset A$ satisfying conditions (32.6.1)-(32.6.3). For any $x \in A$ the sequence of functions (xx_n) converges to x μ-almost everywhere. This sequence is norm bounded and converges to zero pointwise on S. It follows from the Lebesgue theorem on dominated convergence that $\int xx_n d\nu_s \to 0$ and $\int xx_n d\nu_a \to \int x d\nu_a$. Since $\int xx_n d\nu = 0$, then

$$\int x d\nu_a = \lim \int xx_n d\nu_a = \lim \int xx_n d(\nu_a - \nu) = -\lim \int xx_n d\nu_s = 0,$$

whence $\nu_a \in A^{\perp}$ and $\nu_s = \nu - \nu_a \in A^{\perp}$.

32.8. THEOREM. *Let A be a Dirichlet algebra. If $\mu_1, \mu_2 \in \mathfrak{M}(A)$, then either the measures μ_1 and μ_2 are equivalent (mutually absolutely continuous) or they are mutually singular.*

PROOF. Let $\mu_1 = \nu_a + \nu_s$ be the decomposition of the measure μ_1 into the μ_2-absolutely continuous and μ_2-singular parts. We have $\nu_a \geqslant 0$ and $\nu_s \geqslant 0$. If x is any element of the maximal ideal corresponding to the measure μ_1, i.e. an element of A such that $\int x d\mu_1 = 0$, and $x \neq 0$, then the measure $x\mu_1$ is in $A^{\perp}$, since for any $y \in A$ we have

$$\int y d(x\mu_1) = \int yx d\mu_1 = \int y d\mu_1 \int x d\mu_1 = 0.$$

Now, $x\mu_1 = x\nu_a + x\nu_s$ is the decomposition of the measure $x\mu_1$ into the μ_2-absolutely continuous and μ_2-singular parts and by the preceding theorem $x\nu_a \in A^{\perp}$, $x\nu_s \in A^{\perp}$. Let z be any element of A and let f be the multiplicative linear functional represented by μ_1. Then the element $z_0 = z - f(z)e$ is in the maximal ideal corresponding to μ_1, i.e. $z_0 \nu_a \in A^{\perp}$. Hence

$$0 = \int e d(z_0 \nu_a) = \int z d\nu_a - f(z)\nu_a(\Gamma) = \int z d(\nu_a - \nu_a(\Gamma)\mu_1)$$

for all $z \in A$. Since $\nu_a(\Gamma)$ is a real number, then the measure $\nu_a - \nu_a(\Gamma)\mu_1$ is a real measure orthogonal to A. The only such measure is, by assumption, the zero measure, so $\nu_a = \nu_a(\Gamma)\mu_1$. If $\nu_a(\Gamma) = 0$, then $\mu_1 = \nu_s$ and the measure μ_1 is singular with respect to μ_2. If $\nu_a(\Gamma) \neq 0$, then the measure μ_1 is absolutely continuous with respect to μ_2. Similarly the measure μ_2 is either absolutely continuous or singular with respect to μ_1. Thus the two measures either are mutually absolutely continuous or are mutually singular, for a non-zero measure μ never can be absolutely continuous with respect to any μ-singular measure.

We thus see that the set of all multiplicative linear functionals can be decomposed into equivalence classes: two functionals are assigned to the same class if their representing measures are mutually absolutely continuous, in symbols $f_1 \sim f_2$ (or $\mu_1 \sim \mu_2$); the relation $\sim$ clearly is an equivalence relation.

32.9. DEFINITION. Let A be a Dirichlet algebra. The *Gleason parts* of A are those subsets of $\mathfrak{M}(A)$ which are precisely the equivalence classes of the relation $\sim$.

The Gleason part containing an element f will be denoted by $[f]$.

32.9.a. EXERCISE. Find the Gleason parts of the algebra $\mathscr{A}$ (Example 3.9).

32.9.b. EXERCISE. Prove that if $f \in \Gamma(A)$, then $[f] = \{f\}$.

32.10. THEOREM. *Two functionals $f_1, f_2 \in \mathfrak{M}(A)$ are in the same Gleason part if and only if*

(32.10.1) $$\|f_1 - f_2\| < 2.$$

PROOF. Let $f_1, f_2 \in \mathfrak{M}(A)$. Let μ_1 and μ_2 denote the representing probability measures. If $f_1 \notin [f_2]$, then the measures μ_1 and μ_2 are mutually singular, thus concentrated on disjoint sets. Then $\mu_1 - \mu_2$ is the Jordan decomposition of the real measure representing the functional $f_1 - f_2$. Hence we have

$$\|f_1 - f_2\| = |\mu_1 - \mu_2|\,(\Gamma) = \mu_1(\Gamma) + \mu_2(\Gamma) = 2.$$

If $f_1 \in [f_2]$, then $\mu_1 = \varphi\mu_2$, where φ is some non-negative function defined on Γ and belonging to $L_1(\mu_2)$. Moreover,

(32.10.2) $$\int \varphi \, d\mu_2 = \int \varphi \, d\mu_1 = 1.$$

We may assume $\mu_1 \neq \mu_2$. For any $x \in A$, $\|x\| = 1$, we have

$$|f_1(x) - f_2(x)| = \left|\int x(\varphi - 1)\, d\mu_2\right| \leqslant \int |x|\,|\varphi - 1|\, d\mu_2 \leqslant \int |\varphi - 1|\, d\mu_2,$$

so the inequality $\|f_1 - f_2\| < 2$ will be proved if we show that $\int |\varphi - 1|\, d\mu_2 < 2$. Write $K = \{p \in \Gamma : \varphi(p) > 1\}$, $U = \Gamma \setminus K$. Suppose that $\mu_2(K) = 0$. Then $0 \leqslant \varphi \leqslant 1$ μ_2-almost everywhere, whence, in view of (32.10.2), $\varphi = 1$ μ_2-almost everywhere and $\mu_1 = \mu_2$, contrary to the assumption. Thus $\mu_2(K) > 0$, or $\int_K d\mu_2 > 0$ and we may write

$$\int |\varphi - 1|\, d\mu_2 = \int_K (\varphi - 1)\, d\mu_2 + \int_U (1 - \varphi)\, d\mu_2 = \int_K \varphi\, d\mu_2 - \int_U \varphi\, d\mu_2 - \int_K d\mu_2 + \int_U d\mu_2$$
$$= 1 - 2\int_U \varphi\, d\mu_2 + 1 - 2\int_K d\mu_2 = 2\Big(1 - \int_U \varphi\, d\mu_2 - \int_K d\mu_2\Big) < 2.$$

If $\nu \in [C(\Gamma)]'$, $\mu_1, \mu_2 \in \mathfrak{M}(A)$, $\mu_1 \sim \mu_2$, and if $\nu = \nu_a^1+\nu_s^1 = \nu_a^2+\nu_s^2$ are the decompositions of ν into absolutely continuous and singular parts with respect to measures μ_1 and μ_2, then $\nu_a^1 = \nu_a^2$ and $\nu_s^1 = \nu_s^2$. Thus the decompositions with respect to measures belonging to the same Gleason part coincide.

32.11. REMARK. Gleason's original definition of parts started from condition (32.10.1) which was shown to be an equivalence relation; with such a definition the notion of Gleason parts makes a sense in any, not necessarily Dirichlet function algebra.

32.12. DEFINITION. A measure $\sigma \in [C(\Gamma)]'$ is said to be *absolutely singular with respect to the function algebra A* if it is singular with respect to every measure $\mu \in \mathfrak{M}(A)$.

We now are in a position to formulate the announced theorem on the structure of measures belonging to $A^\perp$.

32.13. THEOREM. *Let A be a Dirichlet algebra. Every measure $\mu \in A^\perp$ has the form*

$$\mu = \sum_{i=1}^{\infty} \varphi_i \mu_i + \sigma, \tag{32.13.1}$$

where $\mu_i \in \mathfrak{M}(A)$, $\varphi_i \in L_1(\mu_i)$ *are such that* $\int \varphi_i d\mu_i = 0$ (*i.e.* $\varphi_i \mu_i \in A^\perp$), $i = 1, 2, \ldots, \sum_{i=1}^{\infty} \|\varphi_i \mu_i\| < \infty$ *and the measure* $\sigma \in A^\perp$ *is absolutely singular with respect to the algebra A.*

PROOF. Let $G(A)$ denote the family of all Gleason parts of A and let $\mu \in A^\perp$. If $[\mu_1], \ldots, [\mu_k] \in G(A)$, then write

$$\varrho = \mu - \sum_{i=1}^{k} \nu_a^i,$$

where $\mu = \nu_a^i + \nu_s^i$ is the decomposition of μ into the μ_i-absolutely continuous and μ_i-singular parts (this decomposition depends only on the Gleason parts in question and not on the particular choice of the representatives μ_i). We also may write

$$\varrho = \nu_s^1 - \sum_{i=2}^{k} \nu_a^i,$$

whence it follows that the measure ϱ is μ_1-singular since all μ_i's, thus also all ν_a^i's, $i = 2, \ldots, k$, are μ_1-singular. Similarly ϱ is singular with respect to $\mu_2, \mu_3, \ldots$ $\ldots, \mu_k$, thus ϱ is singular with respect to the measure $\sum_{i=1}^{k} \nu_a^i$. Since for any two mutually singular measures $\alpha, \beta \in [C(X)]'$ we have $\|\alpha+\beta\| = \|\alpha\|+\|\beta\|$, then

$$\|\mu\| = \|\varrho + \sum_{i=1}^{k} \nu_a^i\| = \|\varrho\| + \|\sum_{i=1}^{k} \nu_a^i\| = \|\varrho\| + \sum_{i=1}^{k} \|\nu_a^i\|,$$

whence $\sum_{i=1}^{k} \|\nu_a^i\| \leqslant \|\mu\|$. It hence follows that there are at most countably many Gleason parts $[\mu_i]$ with $\|\nu_a^i\| > 0$, and that $\sum_{i=1}^{\infty} \|\nu_a^i\| \leqslant \|\mu\|$.

Now put

$$\sigma = \mu - \sum_{i=1}^{\infty} \nu_a^i.$$

Since $\mu \in A^{\perp}$, then by Theorem 32.7, all measures ν_a^i are in $A^{\perp}$, hence also $\sigma \in A^{\perp}$. Let $\mu_0 \in \mathfrak{M}(A)$. If ν_a^0, the μ_0-absolutely continuous part of μ, is different from 0, so $\mu_0 \in [\mu_{i_0}]$ for some i_0 and if $\sigma_a^{i_0}$, $\sigma_s^{i_0}$ denote the μ_{i_0}-absolutely continuous and μ_{i_0}-singular parts of σ, then we get $\sigma_a^{i_0} = \nu_a^{i_0} - \nu_a^{i_0} = 0$; so σ is singular with respect to μ_{i_0}, thus also with respect to μ_0. If $\nu_a^0 = 0$, then for each i the measures μ_0 and μ_i are mutually singular and σ again is μ_0-singular. Since μ_0 was chosen arbitrarily in $\mathfrak{M}(A)$, then σ is absolutely singular with respect to the algebra A. Thus we have $\mu = \sigma + \sum_{i=1}^{\infty} \nu_a^i$ and writing $\nu_a^i = \varphi_i \mu_i$, $\varphi_i \in L_1(\mu_i)$, we obtain the asserted formula (32.13.1).

A reader interested in further properties of Dirichlet algebras and their applications is referred to the paper Wermer [1].

§ 33. ANTISYMMETRIC ALGEBRAS

The following remark will be useful in the definition of an antisymmetric algebra (as before all algebras are assumed to posses unit elements).

33.1. Remark. Let A be a function algebra and (A, X_0) — its concrete realization. If $x \in A$ and $x(t)$ is a real-valued function on X_0, then for any other concrete realization (A, X) of the algebra A the function $x(t)$ is also real valued on X. In fact, in view of (29.2.1) $x(t)$ is real-valued on $\Gamma(A)$, so by (29.7.1) it is real-valued on $\mathfrak{M}(A)$ and again by (29.2.1) it is real-valued on X. Similarly a function $x(t)$ constant on X_0 is constant on any X satisfying (29.2.1).

33.2. Definition. A concrete function algebra (A, X_0) is said to be *antisymmetric* if from the fact that $x(t)$ and its complex conjugate $\overline{x(t)}$ are elements of (A, X_0) it follows that $x(t)$ is a constant function. Or, equivalently, if any real-valued function in (A, X_0) is constant. From Remark 33.1 it follows immediately that if one concrete realization (A, X_0) of a function algebra A is antisymmetric, then so are all other realizations (A, X) and in this case we call A an antisymmetric algebra too.

An example of an antisymmetric algebra is the disc algebra (Example 3.9).

The main goal of this stating is to prove the following *Shilov–Bishop theorem* (Shilov [1], Bishop [2], Glicksberg [1]) showing that any function algebra can be expressed in terms of antisymmetric algebras. It reads as follows:

33.3. THEOREM. *Let A be a function algebra and (A, X) its concrete realiza-'on. Then there is a partition of X into a union of disjoint closed sets K_α,*

$$X = \bigcup K_\alpha \tag{33.3.1}$$

uch that

(i) *if $x \in C(X)$ and for restrictions we have $x|_{K_\alpha} \in A|_{K_\alpha}$ for all α, then $x \in A$.*

(ii) *Each restriction $A_\alpha = A|_{K_\alpha}$ is a (complete) antisymmetric algebra.*

This theorem shows that (A, X) can be obtained by "glueing together" a family (A_α, K_α) of antisymmetric algebras. The proof will be obtained through the fol-owing series of lemmas and definitions. First we define the partition (K_α).

33.4. DEFINITION. Let (A, X) be a concrete function algebra. A subset $S \subset X$ s said to be a *set of antisymmetry for A* if for any $x \in A$ from the fact that x is eal-valued on S it follows that x is constant on S. Obviously any subset of X onsisting of a single point is a set of antisymmetry. It is also easy to see that ' (S_α) is any family of sets of antisymmetry having a non-void intersection, then ne union $\bigcup S_\alpha$ is also a set of antisymmetry, and that the closure of a set of ntisymmetry is again such a set. Thus the union of all sets of antisymmetry ontaining a given point $p \in X$ is a (non-void) maximal set of antisymmetry con-aining p and it is a closed set. This implies that two different maximal sets of ntisymmetry are disjoint. If we denote by (K_α) the family of all maximal sets of ntisymmetry for A we obtain a partition of X as in (33.3.1).

33.4.a. EXERCISE. Prove all statements of Definition 33.4.

33.5. LEMMA. *Let μ be an extreme point of the unit ball in $(A, X)^\perp$* (cf. Defini-ion 32.1); *then the support of μ in X is a set of antisymmetry for A.*

PROOF. Denote by S the support of μ and suppose that for some $x \in A$ the unction $x(t)$ is real-valued on S. We have to show that $x(t)$ is constant on S. ince A contains constants, we may assume $0 < x(t) < 1$ on X and so the measures $x\mu$ and $(1-x)\mu$ are non-zero measures in $C(X)'$. We can now write

$$\mu = \|x\mu\| \frac{x\mu}{\|x\mu\|} + \|(1-x)\mu\| \frac{(1-x)\mu}{\|(1-x)\mu\|}. \tag{33.5.1}$$

By our assumption we have also

$$\|x\mu\| + \|(1-x)\mu\| = \int x\,d|\mu| + \int (1-x)\,d|\mu| = \int d|\mu| = \|\mu\| = 1. \tag{33.5.2}$$

Similarly as in the proof of Theorem 32.8 we have $x\mu \in A^\perp$ and $(1-x)\mu \in A^\perp$, o from (33.5.1), (33.5.2) and the fact that μ is an extreme point of the unit ball n A this implies that $\mu = \dfrac{x\mu}{\|x\mu\|}$ which means that x is constant on the support S of μ.

From this lemma we obtain the proof of (i) of Theorem 33.3. Let $x \in C(X)$ and $x|_K \in A|_K$ for every maximal set of antisymmetry K. If $x \notin A$, then there is a measure μ_0 in $A^\perp$, such that $\int x\,d\mu_0 \neq 0$. Without loss of generality we may assume that $||\mu|| = 1$. But this leads to a contradiction since, by Lemma 33.5, for each extreme point μ of the unit ball in $A^\perp$ we have $\operatorname{supp}\mu \subset K_\alpha$ for some α and so $\int x\,d\mu = 0$. Applying now the Krein–Milman theorem we see that $\int x\,d\mu = 0$ for every μ in the unit ball of $A^\perp$.

As an easy corollary of this result we obtain the classical result formulated already in 1.11.

33.6. THEOREM (Stone–Weierstrass). *Let A be a function algebra and (A, X) its concrete realization. If together with any $x(t)$ the algebra A contains its complex conjugate $\overline{x(t)}$ on X, then $A = C(X)$.*

The proof follows immediately from the fact that in this case the maximal sets of antisymmetry consist of single points and from (i) of Theorem 33.3.

Now we pass to the proof of (ii) of Theorem 33.3. First we need the following modification of the concept of a peak set (cf. Definition 19.9).

33.7. DEFINITION. Let (A, X) be a concrete function algebra. We call $E \subset X$ a *peak set of* A if there is an $x \in A$ with $||x|| = 1$ such that $E = \{t \in X: x(t) = 1\}$. If we take $\frac{1}{2}(x+1)$ instead of x (as e.g. in the proof of Theorem 31.3) we see that $|x(t)| < 1$ for all $t \in X \setminus E$. In this case we say that *x peaks on E*.

33.7.a. EXERCISE. Prove that a finite intersection of peak sets is again a peak set.

33.8. LEMMA. *If (A, X) is a concrete function algebra and $E \subset X$ is an intersection of peak sets of A, then the restriction $A|_E$ is a closed subalgebra of $C(E)$.*

PROOF. Put $kE = \{x \in A: x|E = 0\}$, it is a closed ideal in A (cf. 11.2). The lemma will be proved if we show that $A|_E$ with the supremum norm $||x||_E = \sup_E |x(t)|$ is isometrically isomorphic with the quotient A/kE (cf. 4.4). Denote by $\tilde{x}$ the natural projection of $x \in A$ in A/kE. For any $p \in E$ the functional $x \to x(p)$ is constant on cosets of kE, thus it depends only upon $\tilde{x}$ and so defines a multiplicative linear functional on A/kE. This implies $|x(p)| \leqslant ||\tilde{x}||$, or

$$||x||_E \leqslant ||\tilde{x}||. \tag{33.8.1}$$

For a given $\varepsilon > 0$ and $x \in A$ let V be a neighbourhood of E in X given by $V = \{t \in X: |x(t)| < ||x||_E + \varepsilon\}$. Since E is an intersection of peak sets, then by the compactness of X there is a finite intersection F of peak sets which contains E and is contained in V. By Exercise 33.7.a, F is a peak set itself and so there is an element $y \in A$ such that $y(t) = 1$ for $t \in E$ and $|y(t)| < 1$ otherwise. For every natural number n,

$$y^n x - x \in kE \quad \text{and} \quad \limsup ||y^n x|| \leqslant ||x||_V \leqslant ||x||_E + \varepsilon.$$

This implies $\|\tilde{x}\| \leqslant \|y^n x\| < \|x\|_E + 2\varepsilon$, that holds for large n. Since ε was chosen arbitrarily it follows $\|\tilde{x}\| \leqslant \|x\|_E$, which together with (33.8.1) gives the desired isometry (which is clearly an isomorphism) between $A|_E$ and A/kE.

33.9. LEMMA. *If (A, X) is as above, then any maximal set of antisymmetry $K \subset X$ is an intersection of peak sets.*

PROOF. Let K be a maximal set of antisymmetry of A in X. By the embedding $K \subset \mathfrak{M}(A)$ it is also a set of antisymmetry in $\mathfrak{M}(A)$, so it is contained in some maximal set of antisymmetry $K_1 \subset \mathfrak{M}(A)$ and it is clear that

$$K = K_1 \cap X. \tag{33.9.1}$$

It is now sufficient to show that K_1 is an intersection of peak sets E_α in $\mathfrak{M}(A)$, since the intersections $E_\alpha \cap X$ are peak sets in X and $K = \bigcup_\alpha E_\alpha \cap X$.

Thus we have reduced the problem to the case when $X = \mathfrak{M}(A)$. Let E denote the intersection of all peak sets containing K (e.g. X is such a peak set).

We have to show that $K = E$. Suppose, on the contrary, that there is a point $p_0 \in E \setminus K$. If $\mathfrak{M}(A) = K \cup \{p_0\}$, then, by Theorem 20.2, the characteristic function of K is in A and so K is itself a peak set, which gives a contradiction. If $\mathfrak{M}(A) \neq K \cup \{p_0\}$, we can find an open set V containing K and not equal to K, which does not contain p_0. Clearly V contains no peak set containing K, so if we find in V such a peak set, we obtain a contradiction completing the proof. Since $K \nsubseteq V$ and K is a maximal set of antisymmetry, there is in A a real-valued function $x(p)$ which is non-constant on V. Let t_0 denote the constant value of x on K and choose $\delta > 0$ in such a way that

$$t_0 - \delta < x(p) < t_0 + \delta \tag{33.9.2}$$

for all $p \in V$. Put $p(t) = 1-(t-t_0)^2\delta^{-2}$ and $y = p(x)$, so $y \in A$. The function y is non-constant on V, $0 \leqslant y(p) \leqslant 1$ for all $p \in V$ and $y(p) = 1$ for all $p \in K$. Hence there is in V a local peak set containing K (cf. Definition 19.9), which, by Theorem 19.10, is a peak set for A.

Now we can give the proof of (ii) of Theorem 33.3.

By Lemmas 33.8 and 33.9 all restrictions $A_\alpha = A|_{K_\alpha}$ are complete algebras, while by the definitions of the sets of antisymmetry all A_α's are antisymmetric algebras.

Theorem 33.3 reduces in a sense the investigation of function algebras to that of antisymmetric algebras. However, the problem of description of antisymmetric algebras remains. In the paper [2] Gelfand posed several problems on antisymmetric algebras hoping that the positive answers would permit the description of antisymmetric algebras and thus all function algebras in a simple way. Here we list the problems:

33.10. PROBLEMS.

33.10.1. Does there exist an antisymmetric algebra having a one-dimensional maximal ideal space (i.e. $\mathfrak{M}(A) = [0, 1]$)?

Denote by $A(G)$, where G is a domain in $\boldsymbol{C}$, the algebra of all continuous functions on the closure $\overline{G}$ of G and analytic in G. Call a function algebra A an algebra of type $A(G)$ if it is a subalgebra $A(G)$ for a certain G.

33.10.2. Let A be an antisymmetric algebra such that $\dim\mathfrak{M}(A) = 2$; is A then of type $A(G)$? (It is understood here that $\mathfrak{M}(A) \subset \boldsymbol{C}$.)

33.10.3. Is the algebra $A(G)$ a maximal antisymmetric algebra having G as its maximal ideal space?

33.10.4. Does there exists an antisymmetric algebra A with maximal ideal space homeomorphic either to a two-dimensional sphere or to a two-dimensional torus?

33.10.5. Does there exist an antisymmetric algebra having maximal idea space homeomorphic with the three-dimensional cube?

Concerning antisymmetric algebras with four-dimensional maximal ideal spaces we can mention algebras of type $A(G)$ with $G \subset \boldsymbol{C}^2$; but:

33.10.6. Do there exist other antisymmetric algebras with four-dimensional maximal ideal spaces, than those mentioned above? Is the algebra $A(G)$ the maximal antisymmetric algebra with $\mathfrak{M}(A) = \overline{G}$?

Call a function algebra A an analytic algebra if from the fact that $x \in A$ and x vanishes on an open subset of $\mathfrak{M}(A)$ it follows that $x = 0$.

33.10.7. Are the concepts of analycity and of antisymmetry equivalent?

Answering to some of those questions Hoffmann and Singer wrote the paper [1]. We describe here shortly the content of this paper (for the proofs the reader is referred to the original work).

The authors cannot answer questions 33.10.1 and 33.10.2, they notice, moreover, that they can not tell whether there exists a function algebra A with $\mathfrak{M}(A) = [0, 1]$, which is different from $C\,[0, 1]$. The answers to other questions, if any, are negative. The algebras $A(G)$ turn out to be not maximal antisymmetric algebras with $\mathfrak{M}(A) = \overline{G}$ $(\subset \boldsymbol{C})$, there is an antisymmetric algebra having $\mathfrak{M}(A)$ homeomorphic to the three-dimensional cube as well as with $\mathfrak{M}(A)$ equal to the two-dimensional sphere (Problem 33.10.4 is open in the case when $\mathfrak{M}(A)$ has to be a torus). Also the answer to 33.10.6 is negative (all constructions are rather complicated). Concerning relations between antisymmetry and analyticity we mention the following result (also included in Glicksberg [1]).

33.11. THEOREM. *If A is an analytic algebra, then it is also antisymmetric.*

PROOF. Let $x, y \in A$ and $xy = 0$. If $x(p_0) \neq 0$ for some $p_0 \in \mathfrak{M}(A)$, then

$y(p) = 0$ for all p in a neighbourhood of p_0. Since A is analytic, $y = 0$ and sn there are no divisors of zero in A. Put now $A_0 = \{x \in A: \overline{x(t)} \in A\}$. It is easy to see that A_0 is a subalgebra of A and A is antisymmetric if and only if A_0 consists of scalar multiples of the unit. The latter is equivalent with $\mathfrak{M}(A_0) = \{p\}$, a one point set. But, by the Weierstrass–Stone theorem, $A_0 = C(\mathfrak{M}(A_0))$ and if $\mathfrak{M}(A_0)$ consists of at least two points, then A_0 and consequently A possesses divisors of zero. The conclusion follows now from the first remark that there are no divisors of zero in A.

But, on the other hand, one can have on antisymmetric algebra which is non-analytic (it gives a negative answer to question 33.10.6), as follows from the following

34.11.a. Exercise. Put $G = \{t \in \boldsymbol{C}: |t| < 1\} \cup \{t \in \boldsymbol{C}: |t-2| < 1\}$. Then the algebra $A(G)$ is antisymmetric but not analytic (cf. notation before Problem 33.10.2).

In a short letter to the editor of Usp. Mat. Nauk [3], Gelfand gave the following comment on the results of Hoffmann and Singer. The problem of describing function algebras by means of some "elementary algebras" is still open. It turns out that antisymmetric algebras can be fairly complicated and so Gelfand proposes another candidate for "elementary algebras" namely the above mentioned analytic algebras and proposes to reformulate his previous Problems 33.10 substituting there, whenever it makes a sense, an analytic algebra instead of an antisymmetric algebra.

COMMENTS ON CHAPTER V

The concept of a function algebra makes sense and has interesting models also when the underlying space is not compact and the functions—elements of an algebra in question — are not bounded. The outstanding examples are algebras of analytic functions on domains in $\boldsymbol{C}^n$, or, more generally, on analytic manifolds (as e.g. the example of C.3.3). Such algebras are m-convex B_0-algebras, as described in the comments to Chapters I and II.

We shall not describe here possible generalizations of the results of Chapter V but indicate only some, recently investigated classes of function algebras which are not Banach algebras (and so we have to change enumeration of results which will be here C.V.1, C.V.2 etc.).

C.V.1. Definition. Let Ω be a topological space and let $C(\Omega)$ be the algebra of all continuous functions on Ω, it is an m-convex algebra with compact-open topology, i.e. with the topology introduced by means of seminorms of the form $\|x\|_K = \max_{t \in K} |x(t)|$, where K is a compact subset of Ω. By a *function algebra* (or *uniform algebra*) on Ω we mean any separating closed subalgebra of $C(\Omega)$, containing the constants (this definition corresponds rather to Definition 29.2 of a concrete function algebra. The reader can easily introduce an "abstract definition" corresponding to that of 29.1). If A is a function algebra on Ω and $K \subset \Omega$ is a compact subset, then A_K denotes the (Banach) function algebra obtained by completion in the norm $\|x\|_K$ the algebra of all restrictions of A to K.

C.V.2. Definition. A uniform algebra A on the space Ω is called a *maximum modulus algebra* if for any compact $K \subset \Omega$ we have $\Gamma(A_K) \subset \partial K$, where ∂K is the topological boundary and A_K is defined in C.V.1.

C.V.3. DEFINITION. A uniform algebra A is said to be *Montel* if it is a Montel space as a topological linear space (i.e. if every bounded subset of A is relatively compact in A).

The following theorem of Meyers gives a characterization of the algebra $H(U)$ of all holomorphic functions on an open subset $U \subset \boldsymbol{C}$, with compact-open topology (for proof and details cf. Birtel [2] and Meyers [1]).

C.V.4. THEOREM. *Let A be a uniform algebra on an open $U \subset \boldsymbol{C}$, containing the polynomials and assume $\mathfrak{M}(A) = U$ (also topologically). Then the following conditions are equivalent:*

(i) $A = H(U)$,

(ii) *A is a maximum modulus algebra,*

(iii) *A is a Montel algebra,*

(iv) *A has no local peak points* (definition is the same as that of 19.9 adjusted to the more general situation),

(v) *A has no non-zero M-singular elements* (cf. Definition C.14.4 or rather the characterization C.14.5).

For higher dimensions the above theorem is false. Now we consider the situation when $U = \boldsymbol{C}$.

C.V.5. DEFINITION. A uniform algebra A is called *Liouville* if every non-constant element of A has an unbounded spectrum Birtel [1].

C.V.6. THEOREM (Birtel [1]). *Every singly generated Liouville algebra A of type B_0 is isomorphic with $\mathscr{E}$* (cf. Example C.3.3) *provided there is a sequence of seminorms $\|x\|_i$ giving the topology of A such that $\bigcap_i \partial\sigma_{A_i}(x_0) = \emptyset$* (cf. notation of C.10).

As it was shown in Birtel and Lindberg [1], there exists a Loiuville B_0-algebra A with $\mathfrak{M}(A) = \boldsymbol{C}$, which contains $\mathscr{E}$ as a proper subalgebra. For more details on these algebras cf. the above mentioned papers.

Appendix

SOME GENERALIZATIONS OF BANACH SPACES

We describe here some less popular classes of topological linear spaces and ix some terminology used in our comments.

A.1. DEFINITION. A topological linear space X is called an *F-space* if its topology an be given by means of a complete metric. It can be proved that the topology f an F-space X can be given by means of a pseudonorm $||x||$, called sometimes n *F-norm*, which satisfies the following conditions:

A.1.1. $||x|| \geqslant 0$ and $||x|| = 0$ if and only if $x = 0$.

A.1.2. $||\lambda x|| \leqslant ||x||$ for all $\lambda \in \boldsymbol{C}$ with $|\lambda| \leqslant 1$.

A.1.3. $||x+y|| \leqslant ||x||+||y||$ for all $x, y \in X$.

A.1.4. If $||x_n|| \to 0$, then $||\lambda x_n|| \to 0$ for all $\lambda \in \boldsymbol{C}$.

A.1.5. If $\lambda_n \to 0$, then $||\lambda_n x|| \to 0$ for all $x \in X$.

The distance between x and y in X is given by $||x-y||$ and X is complete in is metric (for details cf. Banach [1], Dunford and Schwartz [1], Schaefer [1]).

A.2. DEFINITION. A subset E of a topological linear space X is said to be *bounded* for any neighbourhood U of the origin in X there is a non-zero scalar λ such at $\lambda E \subset U$. A topological linear space X is said to be *locally bounded* if there a bounded neighbourhood U of zero. In this case the scalar multiplies λU, $\neq 0$ form a basis of neighbourhoods for X and X is a topological metric space. his implies that a locally bounded space X is an F-space provided it is complete, herwise its completion, which is also locally bounded is an F-space.

The following result, called the *Aoki–Rolewicz* (Aoki [1] and Rolewicz [1]) *eorem* gives a characterization of locally bounded spaces.

In order to formulate it we need the following

A.3. DEFINITION. An F-norm $||x||$ in the space X is called a *p-homogeneous eudonorm*, $0 < p \leqslant 1$, if $||\lambda x|| = |\lambda|^p ||x||$ for all $x \in X$ and all (real or complex) alars λ.

A.4. THEOREM. *A topological linear space X is locally bounded if and only if topology can be given by means of a p-homogenous pseudonorm* (thus it is an space, or a dense subspace of an F-space).

The basic properties of locally bounded spaces can be found in Köthe [1] where the reader can find also the proof of the following result of Kolmogoroff

A.5. THEOREM. *A topological linear space is a normed space if and only if it is both a locally convex space and a locally bounded space.*

Thus there are two natural generalizations of normed spaces: locally bounded spaces and locally convex spaces. Our comments are devoted mainly to two generalizations of Banach algebras induced in this way.

Let us list some (negative) properties of non-locally convex spaces, which can point out some technical difficulties occurring when dealing with these spaces:

A.6.1. A non-locally convex space (in particular a locally bounded space) may possess no continuous linear functionals.

A.6.2. If X is a non-locally convex space, then there is always a continuous function from the unit interval to X which is not Riemann integrable (cf. Mazur and Orlicz [1]).

We shall make use of the following

A.7. DEFINITION. A locally convex F-space is called a *space of type* B_0, or a B_0*-space*.

Its topology can be given by means of a sequence of (homogenuous) seminorms $\|x\|_1, \|x\|_2, \ldots$ and without loss of generality it may be assumed that $\|x\|_i \leqslant \|x\|_{i+1}$ for all $x \in A$, $i = 1, 2, \ldots$

Under this assumption if f is a continuous linear functional in a B_0-space X then there is a constant $C > 0$ and an index i such that

$$|f(x)| \leqslant C\|x\|_i$$

for all $x \in X$. Similarly if X and Y are B_0-spaces with non-decreasing systems of seminorms $\|\cdot\|_i$ resp. $|||\cdot|||_i$ and if T is a continuous linear map from X into Y, then for any index i there is an index $j(i)$ and a constant C_i such that

$$\|Tx\|_i \leqslant C_i\|x\|_{j(i)}$$

for all $x \in X$, $i = 1, 2, \ldots$ (cf. Mazur and Orlicz [2], and Schaefer [1]).

References

Allan, G. R.

[1] A spectral theory for locally convex algebras, *Proc. London Math. Soc.* (3) 15 (1965), 399–421.

[2] On a class of locally convex algebras, *ibidem* (3) 15 (1967), 91–114.

[3] Some aspects of the theory of commutative Banach algebras and holomorphic functions of several complex variables, *Bull. London Math. Soc.* 3 (1971), 1–17.

[4] *Embedding of the algebra of formal power series in a Banach algebra* (preprint 1971).

Allan, G. R., Dales, H. G., McClure, J. P.

[1] Pseudo-Banach algebras, *Studia Math.* 40 (1971), 55–69.

Aoki, T.

[1] Locally bounded linear topological spaces, *Proc. Imp. Acad. Tokyo* 18 (1942), 588–597.

Arens, R.

[1] The space L_p and convex topological rings, *Bull. Amer. Math. Soc.* 52 (1946), 931–935.

[2] Linear topological division algebras, *ibidem* 53 (1947), 623–630.

[3] The adjoint of bilinear operation, *Proc. Amer. Math. Soc.* 2 (1951), 839–848.

[4] A generalization of normed rings, *Pacific J. Math.* 2 (1952), 455–471.

[5] Dense inverse limit rings, *Michigan Math. J.* 5 (1958), 166–182.

[6] Inverse producing extensions of Banach algebras, *Trans. Amer. Math. Soc.* 88 (1958), 536–548.

[7] Extensions of Banach algebras, *Pacific J. Math.* 10 (1960), 1–16.

[8] Analytic functional calculus in commutative topological algebras, *ibidem* 11 (1961), 405–429.

[9] Ideals in Banach algebra extensions, *Studia Math.* 31 (1968), 29–34.

Arens, R. Calderón, A. P.

[1] Analytic functions of several Banach algebra elements, *Ann. of Math.* 62 (1955), 204–216.

Banach, S.

[1] *Théorie des opérations linéaires*, Warszawa 1932.

[2] Remarques sur les groupes et corps métriques, *Studia Math.* 10 (1948), 178–181.

Birtel, F.

[1] Singly generated Liouville algebras, *Michigan Math. J.* 11 (1964), 84–94.

[2] *Function algebras*, Lecture notes, Univ. of Nijmegen 1968/69.

Birtel, F. (editor)

[1] *Function algebras*, Proc. International Symposium on Function algebras held at Tulane University 1965, Chicago 1966.

Birtel, F., Lindberg, J. A.

[1] A Liouville algebra of non-entire functions, *Studia Math.* 25 (1964), 27–31.

Bishop, E. A.

[1] A minimal boundary for function algebras, *Pacific J. Math.* 9 (1959), 629–642.

[2] A generalization of Stone–Weierstrass theorem, *ibidem* 11 (1961), 777–783.

Bohnenblust, H. F., Karlin, S.

[1] Geometrical properties of the unit sphere of Banach algebras, *Ann. of Math.* 62 (1955), 217–229.

Bonnard, M.

[1] Sur le calcul fonctionnel holomorphe multiforme dans les algèbres topologiques, *Ann. Sci. École Norm. Sup. 4e Serie*, t. 2 (1969), 397–422.

Borel, E.

[1] Sur les zéros des fonctions entières, *Acta Math.* 20 (1897), 357–396.

Bourbaki, N.

[1] *Espaces vectoriels topologiques*, Paris 1953.

[2] *Théories spectrales*, Chapitres 1 et 2, Paris 1967.

Brooks, R. M.

[1] A ring of analytic functions, *Studia Math.* 24 (1964), 191–210.

[2] On locally m-convex *-algebras, *Pacific J. Math.* 23 (1967), 5–23.

[3] Boundaries for locally m-convex algebras, *Duke Math. J.* 34 (1967), 103–116.

[4] On spectrum of finitely generated locally m-convex algebras, *Studia Math.* 29 (1968), 143–150.

[5] A ring of analytic functions II, *ibidem* 39 (1971), 199–208.

[6] Boundaries for natural systems, *Indiana Univ. Math. J.* 20 (1971), 865–875.

Browder, A.

[1] *Introduction to function algebras*, New York 1969.

Cohen, P. J.

[1] Factorization in group algebras, *Duke Math. J.* 26 (1959), 199–205.

Craw, I. G.

[1] Factorization in Fréchet algebras, *J. London Math. Soc.* 44 (1969), 607–611.

[2] A condition equivalent to the continuity of characters on a Fréchet algebra, *Proc. London Math. Soc.* (3) 22 (1971), 452–464.

Dales, H. G.

[1] Boundaries and peak points for Banach function algebras, *ibidem* (3) 22 (1971), 121–136.

Diximier, J.

[1] *Les C*-algèbres et leurs représentations*, Paris 1964.

[2] *Les algèbre d'opérateurs dans l'espace Hilbertien* (*algèbres de von Neumann*), Paris 1969.

Dixon, P. G.

[1] *Generalized* **B***-*algebras*, Ph. D. dissertation, Univ. of Cambridge 1970.

[2] Unbounded operator algebras, *Proc. London Math. Soc.* 23 (1971), 53–69.

[3] An embedding theorem for commutative B_0-algebras, *Studia Math.* 41 (1972), 163–168.

Dixon, P. G., Fremlin, D. H.

[1] *A remark concerning multiplicative functionals on LMC algebras* (preprint Dept. Math. Univ. of Cambridge).

Dunford, N., Schwartz, J.

[1] *Linear operators I*, New York 1958.

[2] *Linear operators II*, New York 1963.

Engelking, R.
[1] *Outline of general topology*, Amsterdam 1968.

Eidelheit, M.
[1] On isomorphisms of rings of linear operators, *Studia Math.* 9 (1948), 97–105.

Gelbaum, B. R.
[1] Tensor products of Banach algebras, *Canad. J. Math.* 11 (1959), 297–310.
[2] Note on the tensor product of Banach algebras, *Proc. Amer. Math. Soc.* 12 (1961), 750–757.
[3] Tensor products and related questions, *Trans. Amer. Math. Soc.* 103 (1962), 525–548.

Gelfand, I. M.
[1] Normierte Ringe, *Mat. Sb.* 9 (1941), 3–24.
[2] On subrings of the ring of continuous functions (russ.), *ibidem* 12 (1957), 249–251.
[3] Concerning a paper of K. Hoffman and I. M. Singer (russ.), *ibidem* 15 (1960), 239–240.

Gelfand, I. M., Raikov, D. A., Shilov, G. E.
[1] *Commutative normed rings*, New York 1964.

Gil de Lamadrid, J.
[1] Uniform cross norms and tensor products of Banach algebras, *Bull. Amer. Math. Soc.* 69 (1963), 797–803.

Gleason, A.
[1] Function algebras, *Sem. analytic functions* 2 (1958), 213–226.
[2] A characterization of maximal ideals, *J. Analyse Math.* 19 (1967), 171–172.

Glicksberg, I.
[1] Measures orthogonal to algebras and sets of antisymmetry, *Trans. Amer. Math. Soc.* 105 (1962), 415–435.

Gramsch, B.
[1] Integration und holomorphe Funktionen in lokalbeschränkten Räumen, *Math. Ann.* 162 (1965), 190–210.
[2] Funktionalkalkül mehrer komplexer Veränderlichen in lokalbeschränkten Algebren, *ibidem* 174 (1967), 311–344.
[3] Die Klasse metrischer linearer Räume L_{Φ}, *ibidem* 171 (1967), 61–78.
[4] Tensorprodukte und Integration vektorwertiger Funktionen, *Math. Z.* 100 (1967), 105–122.

Gunning, R. C., Rossi, H.
[1] *Anayltic functions of several complex variables*, New York 1965.

Hewitt, E., Stromberg, K.
[1] *Real and abstract analysis. A modern treatment of the theory of functions of real variable*, New York 1965.

Hille, E., Philips, R. S.
[1] *Functional analysis and semi-groups*, Providence 1957.

Hirschfeld, R. A., Żelazko, W.
[1] On spectral norm Banach algebras, *Bull. Acad. Polon. Sci.* 16 (1968), 195–199.

Hoffman, K., Singer, I. M.
[1] On certain problems of Gelfand (russ.), *Usp. Mat. Nauk* 14 (1959), 99–114.

Johnson, B. E.
[1] The uniqueness of the (complete) norm topology, *Bull. Amer. Math. Soc.* 73 (1967), 537–539.

Kahane, J.-P., Żelazko, W.

[1] A characterization of maximal ideas in commutative Banach algebras, *Studia Math.* 29 (1968), 339–343.

Kallin, E.

[1] A non-local function algebra, *Proc. Nat. Acad. Sci.* 49 (1963), 821–824.

Kitainik, L. M.

[1] Almost nilpotent elements in commutative B_0-algebras (russ.), *Vest. Mosk. Univ.* 6 (1969), 69–72.

Koosis, P.

[1] Sur un théoreme de Paul Cohen, *C. R. Acad. Sci. Paris* 259 (1964), 1380–1382.

Köthe, G.

[1] *Topological vector spaces I*, Berlin 1969.

Kuczma, M. E.

[1] On a problem of E. Michael concerning topological divisors of zero, *Colloq. Math.* 19 (1968), 295–299.

Le Page, C.

[1] Sur quelques conditions entraînant la commutativité dans les algèbres de Banach, *C. R. Acad. Sci. Paris* 265 (1967), 235–237.

Mallios, A.

[1] On the spectrum of a topological tensor product of locally convex algebras, *Math. Ann* 154 (1964), 171–180.

Mazur, S.

[1] Sur les anneaux linéaires, *C. R. Acad. Sci. Paris* 207 (1938), 1025–1027.

Mazur, S., Orlicz, W.

[1] Sur les espaces métriques linéaires, I, *Studia Math.* 10 (1948), 50–68.

[2] Sur les espaces métriques linéaires, II, *ibidem* 13 (1953), 137–179.

Mazur, S., Ulam, S.

[1] Sur les transformations isométriques d'espaces vectoriels normés, *C. R. Acad. Sci. Paris* 194 (1932), 946–948.

Meyers, W. E.

[1] Montel algebras on the plane, *Canad. J. Math.* 22 (1970), 116–122.

Michael, E.

[1] Locally multiplicatively-convex topological algebras, *Mem. Amer. Math. Soc.* 11 (1952), 49.

Mitiagin, B. S., Edelstein, J. S.

[1] Homotopy type of linear group of two classes of Banach spaces, *Funkcional. Anal. i Priložen.* t. IV, w. III (1970), 61–72.

Mitiagin, B. S., Rolewicz, S., Żelazko, W.

[1] Entire functions in B_0-algebras, *Studia Math.* 21 (1962), 291–306.

Murphy, I. S.

[1] Continuity of positive linear functionals on Banach *-algebras, *Bull. London Math. Sci.* 1 (1969), 171–173.

Nagasawa, M.

[1] Isomorphisms between commutative Banach algebras with an application to rings of analytic functions, *Kōdai Math. Sem. Rep.* 11 (1959), 182–188.

Naimark, M. A.

[1] *Normed rings*, New York 1964.

Neubauer, G.

[1] Zur Spektraltheorie in lokalkonvexen Algebren I, *Math. Ann.* 142 (1961), 131–134.

[2] Zur Spektraltheorie in lokalkonvexen Algebren II, *ibidem* 143 (1961), 251–263.

Phelps, R. R.

[1] *Lectures on Choquet theorem*, Princeton 1966.

Pontriagin, L. S.

[1] *Topological groups* (russ.), Moscow 1954.

Przeworska-Rolewicz, D. Rolewicz, S.

[1] On integrals of functions with values in linearly metric complete spaces, *Studia Math.* 26 (1965), 121–131.

Rickart, C. E.

[1] *General theory of Banach algebras*, Princeton 1960.

[2] Holomorphic convexity for general function algebras, *Canad. J. Math.* 20 (1968), 272–290.

Rolewicz, S.

[1] On a certain class of topological linear spaces, *Bull. Acad. Polon. Sci.* 5 (1957), 479–484.

[2] Example of a semisimple m-convex B_0-algebra which is not a projective limit of Banach algebras, *ibidem* 11 (1963), 459–462.

[3] Entire functions in B_0-algebras concerning dense division algebras, *Studia. Math.* 23 (1963), 181–187.

[4] Some remarks on radicals in commutative B_0-algebras, *Bull. Acad. Polon. Sci.* 15 (1967), 153–155.

Rolewicz, S., Żelazko, W.

[1] Some problems concerning B_0-algebras, *Tensor* 13 (1963), 265–279.

Rossi, H.

[1] The local maximum modulus principle, *Ann. of Math.* (2) 74 (1961), 470–493.

Sakai, S.

[1] *The theory of* W*-*algebras*, Yale University Notes 1962.

Sawoń, Z.

[1] The sets of convergence of power series in B_0-algebras, *Studia Math.* 30 (1968), 135–140.

Schaefer, H. H.

[1] *Topological vector spaces*, New York 1966.

Shilov, G. E.

[1] On rings of functions with uniform convergence (russ.), *Ukrain. Mat. Ž.* III (4) (1951), 404–411.

Sierpiński, W.

[1] Sur l'équation fonctionnelle $f(x+y) = f(x)+f(y)$, *Fund. Math.* 1 (1920), 116–122.

Suciu, I.

[1] Eine natürliche Erweiterung der kommutativen Banachalgebra, *Rev. Math. Pures Appl.* (*Bucarest*) 7 (1962), 483–491.

Tao-Hsing, Hsia (Do-Shing, Sya)

[1] On seminormed rings with an involution (russ.), *Dokl. Akad. Nauk SSSR* 126 (6) (1959), 1223–1225.

[2] On seminormed rings with an involution (russ.), *Izv. Akad. Nauk SSSR, Ser. Mat.* (23) No. 4 (1959), 509–528.

Titchmarch, E. C.

[1] *Theory of functions,* Oxford 1952.

Turpin, Ph.

[1] Sur une classe d'algèbres topologiques, *C. R. Acad. Sci. Paris* 263 (1966), A436–A439.

[2] *Sur une classe d'algèbres topologiques généralisant les algèbres localement bornées,* These, Univ. de Grenoble 1966.

[3] Une remarque sur les algèbres à inverse continu, *C. R. Acad. Sci. Paris* 270 (1970), A1686–A1689.

Turpin, Ph., Waelbroeck, L.

[1] Intégration et fonctions holomorphes dans les espaces localement pseudoconvexes, *C.R. Acad. Sci. Paris* 267 (1968), A160–A162.

[2] Algèbres localement pseudoconvexes à inverse continu, *ibidem* 267 (1968), A194–A195.

Varopoulos, N. Th.

[1] Sur les formes positives d'une algèbre de Banach, *C. R. Acad. Sci. Paris* 258 (1964), 2465–2467.

[2] Algèbres tensorielles et applications a l'analyse harmonique, *Summer School on Topological Algebra Theory, Burges 1966.*

[3] Tensor algebra and harmonic analysis, *Acta Math.* 119 (1967), 51–112.

Waelbroeck, L.

[1] Le calcul symbolique dans les algèbres commutatives, *C. R. Acad. Sci. Paris* 238 (1954), 556–558.

[2] Les algèbres à inverse continu, *ibidem* 238 (1954), 640–641.

[3] Le calcul symbolique dans les algèbres commutatives, *J. Math. Pures Appl.* (9) 33 (1954), 147–186.

[4] Calcul symbolique et ensembles bornèes de fonctions rationelles, *Acad. Roy. Belg. Cl. Sci. Mém. Coll.* (5) 43 (1957), 114–123.

[5] Algèbres commutatives: éléments réguliers, *Bull. Soc. Math. Belg.* 9 (1957), 42–49.

[6] Note sur les algèbres du calcul symbolique, *J. Math. Pures Appl.* (9) 37 (1958), 41–44.

[7] Etude spectrale des algèbres complètes, *Acad. Roy. Belg. Cl. Sci. Mém. Coll.* in 8° (2) 31, No. 7 (1960), 142.

[8] On the analytic spectrum of Arens, *Pacific J. Math.* 13 (1963), 317–319.

[9] About a spectral theorem, *Function algebras,* Chicago 1965.

[10] Continuous inverse locally pseudoconvex algebras, *Summer School on topological algebra theory, Burges 1966,* 128–185.

Wermer, J.

[1] Banach algebras and analytic functions, *Advances in Math.* 1, Fasc. 1 (1961), 51–102.

Williamson, J. H.

[1] On topologizing of the field $C(t)$, *Proc. Amer. Math. Soc.* 5 (1954), 729–734.

[2] Dirichlet algebras, *Duke Math. J.* 27 (1960), 373–382.

[3] *Seminar über Funktionen-Algebren,* Berlin 1964.

Yood, B.

[1] Faithful *-representations of normed algebras, *Pacific J. Math.* 10 (1960), 345–363.

[2] On axioms for **B***-algebras, *Bull. Amer. Math. Soc.* 76 (1970), 80–82.

Żelazko, W.

[1] On the locally bounded and m-convex topological algebras, *Studia Math.* 19 (1960), 333–356.

[2] A theorem on B_0-division algebras, *Bull. Acad. Polon. Sci.* 8 (1960), 373–375.

[3] On the algebras L_p of locally compact groups, *Colloq. Math.* 8 (1961), 115–120.
[4] A theorem on discrete groups and algebras L_p, *ibidem* 8 (1961), 205–207.
[5] On the radicals of p-normed algebras, *Studia Math.* 21 (1962), 203–206.
[6] Analytic functions in p-normed algebras, *ibidem* 21 (1962), 345–350.
[7] Some remarks on topological algebras, *ibidem* 22 (1963), 141–149.
[8] Metric generalizations of Banach algebras, *Dissertationes Math.* 47 (1965).
[9] On generalized topological divisors of zero in m-convex locally convex algebras, *Studia Math.* 28 (1966), 9–16.
[10] On generalized topological divisors of zero in real m-convex algebras, *ibidem* 28 (1967), 241–244.
[11] On topological divisors of zero in p-normed algebras without unit, *Colloq. Math.* 16 (1967), 231–234.
[12] A characterization of multiplicative linear functionals in complex Banach algebras, *Studia Math.* 30 (1968), 159–161.
[13] A characterization of Šilov boundary in function algebras, *Comment. Math.* 14 (1970), 63–68.
[14] On permanently singular elements in commutative m-convex locally convex algebras, *Studia Math.* 37 (1971), 181–190.
[15] On a certain class of non-removable ideals in Banach algebras, *ibidem* 44 (1972), 87–92.

Index of symbols

Author index

Subject index